Analog and Mixed-Signal Test

Editor

Bapiraju Vinnakota
Department of Electrical and Computer Engineering
University of Minnesota
Minneapolis, MN 55455
bapi@ece.umn.edu

ISBN 0137863101

90000

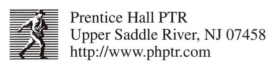

Prentice Hall PTR
Upper Saddle River, NJ 07458
http://www.phptr.com

9 780137 863105

Library of Congress Cataloging-in-Publication Data

```
Analog and mixed-signal test/editor, Bapiraju Vinnakota
     p.  cm,
     Includes bibliographical references and index.
     ISBN 0-13-786310-1
     1. Linear integrated circuits--Testing. 2. Mixed signal
     circuits--testing. I. Vinnakota, Bapiraju.
     TK7874.A543 1998                                    98-4505
     621.3815--dc21                                      CIP
```

Editorial/production supervision: *Nicholas Radhuber*
Cover design: *Scott Weiss*
Cover design director: *Jerry Votta*
Acquisitions editor: *Bernard Goodwin*
Marketing manager: *Miles Williams*
Manufacturing manager: *Alexis R. Heydt*

 © 1998 by Prentice Hall PTR
Prentice-Hall, Inc.
A Simon & Schuster Company
Upper Saddle River, NJ 07458

Prentice Hall books are widely used by corporations and government agencies for training, marketing, and resale.

The publisher offers discounts on this book when ordered in bulk quantities.
For more information, contact Corporate Sales Department, Phone: 800-382-3419;
Fax: 201-236-7141; E-mail: corpsales@prenhall.com
Or write: Prentice Hall PTR, Corp. Sales Dept., One Lake Street, Upper Saddle River, NJ 074

Product names mentioned herein are the trademarks or registered trademarks of their respectiv

Printed in the United States of America
10 9 8 7 6 5 4 3 2 1

ISBN 0-13-786310-1

Prentice-Hall International (UK) Limited, *London*
Prentice-Hall of Australia Pty. Limited, *Sydney*
Prentice-Hall Canada Inc., *Toronto*
Prentice-Hall Hispanoamericana, S.A., *Mexico*
Prentice-Hall of India Private Limited, *New Delhi*
Prentice-Hall of Japan, Inc., *Tokyo*
Simon & Schuster Asia Pte. Ltd., *Singapore*
Editora Prentice-Hall do Brasil, Ltda., *Rio de Janeiro*

Contents

List of Figures

List of Tables

Preface

T here has been a dramatic rise in interest in analog and mixed-signal IC test in the recent past. The increased interest has been motivated by economic factors and advances in technology. Characterizing recent activity as an "explosion of interest" would not be misleading. This book is intended to be a single source reference for recent activity in this area. To meet the needs of its intended audience this book offers several features:

- A diverse range of topics covering all aspects of analog and mixed-signal IC test ranging from fault modeling through built-in self-test are reviewed.
- In addition to reviewing current research, each chapter discusses the advantages and limitations of each potential test solution.
- Virtually every chapter contains applications of test techniques to circuits of practical interest to designers.
- A comprehensive list of references, up to date as of December 1997, are included in each chapter.
- The contributors are distinguished researchers from academe and industry.

This book should be useful to both practitioners and researchers. Commercial circuit designers, test engineers and CAD developers will be able to use the book as a guide to what is available in terms of test solutions. Researchers in the area of design and test can use it as a starting point for their efforts. Most topics have been covered in great detail, with many illustrative examples. The breadth of and the level of detail in the coverage should also permit the material in this book to be used as a textbook for an advanced graduate course on testing.

Chapter 1 outlines current research in analog test and discusses the two significant approaches: specification-based test and fault model based test. The contents of the book are discussed in more detail, and the terminology used is established in this chapter. The chapter examines the applicability of digital test methods to analog circuits and the impact of differences between digital and analog circuits on test techniques and algorithms.

Chapter 2 reviews work in the area of fault modeling. The review focuses on inductive fault analysis. The coverage is detailed and includes the application of inductive fault analysis to several manufactured circuits.

Chapter 3 is a tutorial survey of recent efforts in the area of fault simulation for analog and mixed-signal circuits. A wide range of algorithms are discussed, and illustrated using examples.

Chapter 4 contains a review of efforts in automatic test pattern generation for analog and mixed-signal circuits. The discussion is focused on the intuition behind various approaches, rather than on the specific details of the various approaches.

Chapter 5 discusses a variety of design for test techniques. The discussion is oriented to the types of circuits used commonly by designers such as data converters, filters and operational amplifiers.

Chapter 6 reviews spectrum-based built-in self test methods. The review discusses the methods used to generate signals on chip, observe the results, their relative quality and BIST schemes for circuits of practical interest.

Chapter 7 introduces the new board-level analog test bus standard. In addition to the standard itself, the chapter discusses the application of the standard in example circuits and methods to measure signals of interest.

Chapter 8 covers the class of switched-current circuits. The chapter introduces switched-current structures for common applications, and methods to test these circuits. The topics discussed include fault modeling, test generation and built-in self test schemes.

Every effort has been made to eliminate overlap between chapters and to eliminate errors from the material in this book. Of course, a few more are likely to have escaped the attention of our test process. Any information from readers about the remaining errors will be much appreciated.

Contributors

Ramesh Harjani

Department of Elec. and Comp. Engg.
University of Minnesota
Minneapolis, MN 55455, USA
harjani@ee.umn.edu

Gordon Roberts

Department of Electrical Engineering
McGill University
Montreal, H3A 2A7, Canada
roberts@macs.ee.mcgill.ca

Manoj Sachdev

Department of Electrical Engineering
University of Waterloo
Waterloo, N2L 3G1, Canada
m.sachdev@uwaterloo.ca

C.-J. Richard Shi

Department of Electrical Engineering
University of Iowa
Iowa City, IA 52242, USA
cjshi@eng.uiowa.edu

Mani Soma

Department of Electrical Engineering
University of Washington
Seattle, WA 98195, USA
soma@ee.washington.edu

Steve Sunter

Logic Vision
Ottawa, K1Z 8R9, Canada
sunter@lvision.com

Benoit Veillette

Department of Electrical Engineering
McGill University
Montreal, H3A 2A7, Canada
benoitv@macs.ee.mcgill.ca

Chin-Long Wey

Department of Electrical Engineering
Michigan State University
East Lansing, MI, USA
wey@egr.msu.edu

Introduction

Bapiraju Vinnakota and Ramesh Harjani

Effective test procedures are a necessary component of any high-quality manufacturing process. Errors in manufacture can result in the production of defective units which need to be discarded. The primary goal of a test process is to determine if the units produced by the manufacturing process are defect-free and will function as desired. A second important goal is to generate information that can be used to improve process yield and thereby reduce costs. A perfect test process should fail all unacceptable units and pass all acceptable units. While perfection is a laudable goal, overly high test costs may negate the economic benefits of producing a high-quality product. A practical test process attempts to minimize the number of bad units passed and the number of good units failed at an acceptable cost. Test processes are also often used to bin parts according to the performance they offer. Though the goals of a test process are simple to articulate, the methods to achieve it are not. Systematic test methodologies for digital integrated circuits have been investigated and developed for more than three decades. However, advances in technology, such as increasing integration, continually erode the effectiveness of current techniques. A second source of test difficulties is the increasing commercial prominence of analog and mixed-signal ICs.

Correspondingly, analog and mixed-signal IC test has recently received substantial attention from academic researchers, as well as from industry. If activity is a measure of importance, analog and mixed-signal (AMS) IC test is one of the hottest issues in test research and development. The increase in activity is evidenced by a jump in the number of conference and journal papers (over 800 in the last four years alone), conference sessions (the theme of the 1997 International Test Conference) and workshops, and special journal issues devoted to analog and mixed-signal IC test. A new IEEE standard for analog and mixed-signal test is being actively developed and will be available in the near future.

1.1 Motivation

The growth in interest has been motivated by advances in IC manufacturing technology and by economic factors. ICs with digital, analog, and mixed-signal circuits on the same substrate are common[1]. The applications for such ICs include wireless communication, networking, multimedia, process control, and real-time control systems. In many cases, systems consist of digital cores surrounded by peripheral analog circuitry, such as filters and data converters. The analog and mixed-signal components serve as interfaces between digital processing circuitry and real-world analog signals. The analog component of the IC usually occupies a much smaller fraction of silicon area than the digital component. In micro-electro-mechanical systems (MEMS), mechanical and electromechanical components are integrated with electronic circuits on a single substrate. In MEMS, analog and mixed-signal circuitry will be essential to enable bidirectional interaction between the digital processing circuitry and the (electro-) mechanical components. It is reasonable to expect that for the foreseeable future analog and mixed-signal circuits will be present in a significant number of mass manufactured ASICs. As with digital ICs, methods to adequately test AMS circuits on ICs are naturally of interest.

While simple curiosity may motivate academic research, it is often not a sufficiently enticing motive to spark commercial activity about a specific issue. Economic factors have also played a role in the recent revival of interest in AMS IC test, especially from industry. Even with digital circuits, test costs for modern ICs are a large fraction of manufacturing costs. Some estimate that test costs are as much as a third of total manufacturing costs. Increasing circuit complexity, reduced access, and increasing die sizes are expected to put further pressure on these costs. In recognition of these costs, several CAD tool vendors have introduced tools for test generation, design-for-test, and built-in self-test. Traditionally, test issues were addressed after a design was largely complete. CAD tools for test are now used to integrate test goals with the design process, well before the design is completely specified. Similar design automation tools are needed for AMS ICs. As discussed above, AMS circuits will not necessarily occupy a large fraction of the signal area in an IC. However, without any effective CAD support, the AMS components will consume test resources and accrue costs completely out of proportion to their size. Conversely, without adequate test the AMS components may lead to disproportionate product failure and yield losses. Hence, there is a strong economic imperative to advance analog and mixed-signal IC test.

1.2 History

Traditionally, analog circuits have been tested by verifying whether they meet all their specifications. This is a straightforward method to ensure that a circuit meets the specifications of the design process. Test inputs can be generated from the specifications. Usually test application is expensive since the number of specifications is large. The first efforts in developing alternative test methods for analog circuits occurred in the 1960s. Early research efforts concentrated on discrete analog circuits. This research was coincidental with early efforts on digital circuit test techniques. Discrete circuit components were often unreliable. Since many circuit components failed during operation, it was essential to be able to repair defective circuits. In other words, fault detection, determining if a system was functioning correctly, was not sufficient. Fault diagnosis, identifying the faulty component(s) so that it could be replaced was also important. Test methods focused on computing currents in all the branches and the voltages at all the nodes in the circuit. The emphasis was on computing all branch currents and node voltages by measuring as few branch currents and node voltages as was possible. The computed values were compared to their expected values. The expected values were determined by simulation assuming nominal values for all circuit components. A difference between computed and expected values indicates the presence of a fault. The same information could be used to locate (and replace) a faulty component. For a substantial period, advances in integration caused digital circuit test to be of more interest than analog circuit test. Recent advances in integration for analog circuits revived interest in analog test. For a comprehensive review of earlier efforts in analog test, readers are referred to the book by Liu [2].

1.3 Current Research

This book reviews recent efforts in analog and mixed-signal test. The direction of recent research has been substantially different from that of earlier research. The shift is necessary because of the fact that the types of circuits and the technology targeted by a test process have changed markedly. Research has naturally concentrated on integrated circuit test. Integration increases test difficulties in many ways. As with digital ICs, access to on-chip signals is decreased. When an analog circuit is buried inside a larger system, even its inputs and/or outputs may not be directly accessible. Measuring them accurately is not a simple task. The emphasis is on low-cost manufacturing-time test for product binning, rather than testing in the field. (Built-in self test schemes can obviously be used in the field as well.) Since repair is not possible, fault diagnosis for repair is not a priority in manufacturing-time test, at the IC level. (Fault diagnosis is necessary when the yield of the process is significantly below its expected value. The information generated by diagnosis is used to correct the manufacturing process.) Mixed-signal circuits have also received more attention since they are very likely to be integrated with digital circuits.

1.3.1   Influence of Digital Test

Research in AMS IC test has been strongly influenced by the advances in digital IC test techniques and algorithms. With the benefit of experience, many processes and developments which occurred sequentially in digital test are taking place concurrently in AMS IC test. For example, commercial tools which enable the integration of test concerns into the design process are a relatively recent development for digital ICs. In contrast, efforts have already been made to integrate AMS IC test methods into the design process. The recent flurry of activity has produced a large number of potential test solutions. However, digital test is still far more advanced than is AMS IC test. The primary reason is the absence of a widely accepted paradigm for analog and mixed-signal circuit test.

For digital ICs, the stuck-at fault model combined with output logic level monitoring has been the primary paradigm for a substantial period of time. Nearly all test algorithms and techniques were based on the stuck-at model. Algorithm development could be based on structural descriptions. Algorithms were evaluated on the basis of the fault coverage they provided. Design for test methods were evaluated on the basis of the improvement in fault coverage they provided. Their cost was assessed according to their impact on area and performance. Test sets could be compacted as long as the stuck-at fault coverage was preserved. More recently, IDDQ test methods and circuit-specific models, such as memory error models, were developed to address a perceived gap in real defect coverage. Interestingly, though the stuck-at model was popular for functional test, a model for performance test was not commonly accepted as easily. Several models, such as the transition fault model, the gate-delay fault model, and the path-delay fault model were developed to model the impact of defects on circuit delay. However, even for a metric such as the delay, which is essentially an "analog" parameter, test generation techniques could be based on structural descriptions.

1.3.2   Analog Test Issues

Given the success of structural fault model-based techniques with digital ICs, there has naturally been significant interest in developing similar techniques for AMS ICs. In spite of the similarity of the goals of testing, there are substantial differences between digital and AMS IC test. The main source of test difficulty in digital circuits is the large size of digital circuits, and the corresponding complexity. Size is not a significant limitation in AMS circuits. Rather, the behavior of circuit signals and the flow of information in analog and mixed-signal circuits is substantially different from that in digital circuits. From a test point of view, these differences impact the ability to form fault models, to generate test sets, and to form design-for-test and built-in self-test schemes. Hence, many digital test and DFT techniques cannot be directly extended to analog and mixed-signal circuits. A few examples of the differences will illustrate the unique difficulties associated with AMS-IC test. This discussion is meant to be illustrative and not comprehensive. Each chapter also reviews analog test issues relevant to the material covered in the chapter.

1.3.2.1   Analog Circuit Behavior

An obvious difference between the two types of circuits is that the number of values an analog signal may possess is infinite, though the range is limited. This is true even in switched capacitor circuits. Though the signals are discrete (in time), they are still analog signals. In a digital circuit, the value of an output for a specific input is uniquely defined. The value of a digital signal can be characterized with complete accuracy. Consequently, it is very easy to determine if a measured digital signal is correct or erroneous. This process is considerably more difficult with analog signals. In contrast, in AMS circuits, because signals are multi-valued, a *range* of values about a nominal value is deemed acceptable for an analog signal. Thus, the test process has to determine if the output is within a range of values, not if it is a specific value. The acceptable *tolerances* in signal values are determined by several factors, such as simulation and measurement inaccuracies and process variations.

Simulation/Measurement Accuracy: To determine if a specific signal is erroneous, its measured value has to be compared to its expected value. The expected value of an analog (or any other!) signal is computed using simulation. The accuracy of simulation is limited by the numerical accuracy of the simulation algorithm, as well as the assumptions made in simulating the circuit. For example, accurately modeling and representing parasitics is crucial. Process variations can also cause the expected value of a signal to vary from its nominal value in different instances of a circuit. (The impact of process variations is discussed below in more detail.) In summary, the expected value of a signal can only be computed (through simulation) with finite accuracy.

Similarly, it is of course not possible to measure signals with infinitesimal accuracy. Several factors may cause errors in the measurement process. Some sources of error, such as offsets, the load of the measuring instrument, and its impedance are systematic and their effects can be accounted for. Other sources such as noise are random in nature and it is expensive to compensate for their effect. All of these factors limit the bandwidth and accuracy of measurement techniques. Integration increases measurement difficulties in many ways. For example, in digital ICs, one technique used to observe internal signals is to transport them to an output pin. In an AMS circuit, if done carelessly, this may actually alter the signal and even the circuit's functionality, since the wire used to transport the signal being observed is an additional load. When digital, analog, and mixed-signal circuits are realized on the same substrate, capacitive coupling from rapidly switching digital signals is an additional source of noise during normal operation and measurement.

In summary, both measurement and simulation are limited in their accuracy. The greater of the two uncertainties, in measurement and simulation, will determine the acceptable tolerance ε in a signal being monitored. Usually the value of ε can only be conservatively estimated.

Process Variations: Unlike discrete circuits, components in an IC cannot be tested individually. The performance of an analog circuit is strongly dependent on the manufacturing process. Process variations can have a significant impact on component parameter values. Design and circuit layout techniques to minimize the impact of process variations have been developed and are well known. For example, absolute component values may vary from their expected values by as much as 20%. However, relative matching between components is much better, as low as 0.1%. Hence, a circuit's functionality is designed to depend on ratios of components rather than products or individual component values. Even with good design techniques, parametric variations (caused by process variations) can impact the functionality of a circuit. Digital circuit fault models and test are based on a single (locally-)catastrophic fault model. (The stuck-at model is the most common one, but not the only such model.) The test generation process assumes that defects cause a significant localized impact at a specific point in the circuit. It is also assumed that no more than one modeled fault is present in the circuit under test. Such a model is adequate to represent the impact of the spot defects targeted in functional test, or that of large parametric variations. Secondly, many multiple catastrophic faults are detected by a test vector targeting one of the constituent single faults.

In analog and mixed-signal circuits, multiple minor variations in several components are as important as large variations. Such multiple variations cannot be effectively represented by a localized catastrophic fault model. It is possible for multiple minor variations in several components to cumulatively cause a device to fail to meet its specifications. Unfortunately, it is also possible for the components of a multiple parametric variation to cancel each others' effects such that the functionality of the circuit is not affected. In other words, not every multiple parametric variation is actually a fault! Simulation is usually the only method to determine if a specific multiple parametric variation is a fault. Given a multiple parametric fault[1], individual component variations may not be detectable, or tests for component variations may not be tests for the multiple fault. Thus, multiple parametric variations cannot be modeled using a single local fault model. Yet, techniques to detect multiple parametric faults are essential since they are far more likely to cause an analog or mixed-signal circuit to malfunction than catastrophic faults. Model-based techniques for parametric faults are very limited in their ability. The set of multiple parametric faults cannot be enumerated. Worst-case analysis leads to pessimistic estimates of the fault coverage. Statistical methods are the only effective techniques to assess the impact and fault coverage of multiple parametric faults. Obviously, model-independent test methods, such as specification-based tests, do not suffer from these limitations.

Functionality/Performance: In digital circuits, there is a clear distinction between functionality and performance. The functionality, that is the relationship between input logic values and output logic values, is specified by the truth table. The performance is specified by the delays of the critical paths. This distinction enables one to test functionality and performance separately. It is possible for a circuit to function correctly without meeting performance specifications. Separate fault models can be and have been developed for functionality and performance. Such a separation is not possible for analog circuits. Functionality and performance are closely intertwined in AMS ICs, often, the two cannot be separated. For example, the frequency response characteristic is the response to time-varying input signals. This is a serious obstacle to proving the effectiveness of structure-based fault models. Developing separate models for function and performance is difficult. Where it is possible, in circuits such as data converters, structural fault models have been shown to adequately model faults which affect the I/O relationship. However, circuit structure alone does not determine performance. Altering device sizes while preserving topology can substantially alter performance. Thus, it is far harder to establish a quantitative relationship between a structural fault model and circuit performance.

Information Flow: Unlike most digital circuits, there is no unique direction to information flow in AMS circuits. Some circuits, such as data converters are exceptions. Multidirectional information flow implies that it is difficult to test circuits by individually testing subcircuits. It is possible that a circuit obtained by combining individually acceptable subcircuits may not meet specifications. The converse is also possible. (This is similar to the impact of a multiple parametric fault.) Similarly, reconfiguring circuit components during test, so as to make the resulting structure easily testable, also may not be acceptable. Adequate performance in a reconfigured structure does not guarantee that the original circuit will meet the specifications.

---

1.   The term "parametric fault" is used to refer to two types of faults. In one definition, the phrase refers to the variations in circuit *component* values. This is the version used in this paragraph. In the second definition, the phrase refers to a fault (that could be locally catastrophic) that causes minor variations in an *output* parameter. Both definitions of the term are used in this book. The use will be clear from the context.

1.3.2.2   Analog Circuit Design

Because of the differences in behavior, the procedures used to design AMS ICs are substantially different from those used to design digital ICs. Analog circuit design is much more of an art than a science. An average digital circuit, one that is not high speed, low power, or low area can be realized entirely using design automation tools. This is not true for analog circuits. Several factors make analog circuit design more complex. Consequently, design practices are very "cultural," specific even to a design group. Readers are referred to books on analog IC design for more information [3].

Specifications: In general, each class of analog circuits has a separate set of specifications. The set of performance specifications for a data converter is quite different from the set of specifications for an operational amplifier. There is no universal set of performance specifications for analog circuits. (This is one of the reasons why analog hardware description languages have been harder to develop.) Correspondingly, design procedures and design techniques for analog circuits tend to be circuit-specific. No general design techniques applicable to all analog circuits exist. At best, generalization is restricted to classes of circuits. Since specifications and design procedures are circuit-specific, it is reasonable to speculate if a general analog test methodology can be developed. Test procedures may also have to be specific to each class of circuits.

Level of Abstraction: AMS circuits present several automation difficulties with respect to test. With a few exceptions, digital test has been very resistant to the hierarchy used to simplify the rest of the design process. Most test generation is performed at the logic level. High-level testability analysis measures have not gained common acceptance. The problem is likely to be worse in AMS circuits. Typically, AMS designs are conceived and simulated at the circuit level. Circuit-level simulation is far more expensive than logic level simulation. Any fault simulation algorithms will necessarily be more complex than logic-level fault simulation algorithms. The same is true for other aspects of test generation.

1.3.3   Test Paradigms

The unfortunate implication of these differences is that successful digital circuit test methods cannot easily be ported to AMS ICs. Even the goals of the test process are affected. The functional test of a digital circuit classifies it as either good or bad. AMS IC test is closer to performance test for digital systems. The test process is likely to be used to bin products, rather than simply classify them as being good or bad. The unique characteristics of analog ICs have made it difficult to formulate a test paradigm that is commonly accepted. This is a serious limitation since it implies there is no common benchmark by which all test techniques can be evaluated. When comparing two test techniques it is difficult to compare their merits and limitations when they have not been referenced to a common set of metrics.

Specification-based tests offer the advantage of ensuring that a circuit which passes the test process will meet a user's needs. Test generation is straightforward since tests can be generated directly from the specifications. Further, such an approach can adapt easily to the different types of circuits. The disadvantage is that with a large number of specifications, test application can be very expensive. Test set size can be reduced by identifying dependencies between specifications. Then, the test inputs need only consist of the set of independent specifications. While this approach reduces test set size, it does not lower the burden of accurately applying and measuring a wide range of test inputs. One solution to this problem is to add on-chip circuitry which can generate test inputs and measure circuit responses. This reduces the burden on external testers.

With structural fault model-based test, test sets are designed to target a specific set of modeled faults. A fault model-based test process offers several enticing advantages. The quality of any set of potential test vectors can be easily quantified in terms of the fault coverage they provide. In other words, test sets can be *graded*. Valid models provide a method to reduce test set size without impacting test quality. Test inputs which do not detect any faults not already detected by other inputs in the test set can be discarded. Models also enable the formulation of design for test schemes which aim to improve fault coverage and/or reduce test costs by altering a circuit. Self-test schemes based on fault models have also been proposed. Model-based test suffers from one very important limitation. No proponents of structural fault models have been able to establish a satisfactory formal link between fault detection and the satisfaction of design specifications. Consequently, there has been great reluctance to accept structural test methods. Equally importantly, since no single fault model has been accepted by all, the various model-based techniques cannot easily be compared.

In summary, the sources of test complexity in analog and mixed-signal circuits are different from those in digital circuits. This implies test techniques may have to be substantially different. Unfortunately, there is no consensus on the basic approach to be followed.

1.4 Organization

This book reviews recent efforts in analog and mixed-signal IC test. The organization of this book owes much to the textbook on digital testing by Abramovici, Breuer, and Friedman[4]. The topics covered recognize the diversity of opinion in this field. Efforts in both specification-based and (structural) fault model-based test are reviewed. The different contributors in this book come down squarely on both sides of the issue. The reader is invited to make a reasoned judgment based on the arguments presented by the contributors.

Fault Modeling   An effective fault model is a fundamental prerequisite for a successful model-based test strategy. The fault list is the set of all modeled faults and the tests generated by the test process are designed to detect all modeled faults. Ideally, the generated test set should also detect any possible physical defects that may occur. It is possible, in theory, for a fault model to be effective without reflecting reality. Thus, a model-based test set may be indirectly effective at detecting physical defects. It is generally accepted that the stuck-at fault model does not reflect reality in digital ICs. Yet, stuck-at model-based test sets have provided a reasonable level of defect coverage in many ICs. In an effort to replicate the success of the stuck-at model, many researchers have developed models which derive a fault list from the structure of a circuit. The computational effort needed to form a fault list is small. The layout of the circuit and the fabrication technology are not used to generate a list of faults. Only structural descriptions are used to generate fault lists. As mentioned above, since structure alone does not define performance, establishing the validity of such a fault model, without actual use in a manufacturing test process, is a difficult task.

It has been suggested that a fault model that reflects reality, that represents the way defects alter a circuit, is likely to be effective. Inductive fault analysis (IFA) attempts to do this by using information about structure of a circuit, its layout, and fabrication technology to generate fault lists. IFA models the impact of process variations and other potential manufacturing defects. Process statistics are used to estimate the probabilities of the occurrence of various defects. Simulation is used to model the impact of defects at the circuit level, and generate a fault list ordered according to the probability of occurrence. While they are closer to physical reality, IFA models are extremely expensive to generate. Prof. Manoj Sachdev, formerly of Philips Research Laboratories, reviews recent efforts in the IFA-based fault model generation. An important aspect of the research reported is that the methods discussed are validated on a real manufacturing process used by Philips.

Fault Simulation/Test Generation   The lack of a fault model has not deterred work in developing test generation and fault simulation algorithms for AMS ICs. Perhaps the lack of a fault model has actually created an environment that encourages a diverse range of ideas. As mentioned above, some test methods use the circuit specifications to derive test inputs. This approach usually leads to a large number of tests. Some researchers have addressed this issue by using linear circuit models about operating points. The linear models are used to identify the number of independent parameters affecting the input/output relationship. Matrix triangularization techniques can be used to identify the number of independent parameters. This information is in turn used to reduce the number of tests, and order them so as to minimize test application time.

Many authors have suggested a model-based test generation/fault simulation process to generate test sets. Several transistor-level fault models have been developed in this context. This process works as follows. The fault model is used to generate a fault list. Faults on the list are processed individually by a test generator, and deleted when detected. Every test generated is simulated with the remaining undetected faults to verify if any of them are coincidentally detected. The transistor level simulation needed for test generation and fault simulation is very expensive. Researchers have suggested the use of transform techniques, hierarchical modeling and behavioral simulation. Many test generation techniques are also targeted at specific types of circuits. Unorthodox test generation methods such as monitoring power supply currents, and ramping supply current, have also been suggested. While test generation methods provide good coverage for modeled faults, as discussed earlier, the link between fault coverage and the satisfaction of specifications has not been satisfactorily established. Professor Richard Shi reviews recent research in fault simulation for analog and mixed-signal ICs and Professor Mani Soma reviews advances in test generation techniques for AMS-ICs.

Design for Test   As digital circuit size and complexity increased, algorithmic improvements alone could not provide an adequate fault coverage. Design for test methods which alter a circuit to improve its testability were developed in response. Scan is the most widely known DFT scheme. In contrast to digital circuits, the development of DFT methods in AMS ICs has paralleled research in fault models and test algorithms. A circuit's testability can be increased in many ways. Initially, methods developed for digital circuits were extended to AMS ICs. One example is that of increasing access to internal signals to improve the ability to control and observe them. Analog scan chains which enable several internal signals to be observed have also been developed. Another method that has been extensively investigated and used in practice is that of partitioning a large circuit into smaller subcircuits, each of which is individually tested. DFT methods developed for digital ICs were designed to address those factors that made digital circuits difficult to test. The same factors may not be as important in AMS ICs.

Several methods specific to AMS ICs have been suggested recently. Several deal with the difficulty of measuring analog signals. One popular method has been to use codes which impose relationships between on-chip signals. The codes are chosen such that they are easy to verify using on-chip checkers which produce digital outputs. Examples include data duplication (in time and/or space) and checksum codes. A second approach is to alter the structure of a circuit during test, or to improve fault coverage with a specific test technique. For example, it has been suggested that filters be reconfigured as oscillators during circuit test. As with other phases of test, the lack of a common paradigm makes it difficult to validate the real quality of a DFT scheme. Indeed many fault models have been developed by authors of papers specifically to validate their DFT technique. Fortunately, most DFT schemes are not tied to specific fault models. Recent research in this area is reviewed by the editor and Professor Ramesh Harjani.

Built-in Self-Test   Increasing package sizes and circuit performance have been driving digital tester costs rapidly upward. Built-in self-test methods aim to control rising tester costs by moving test stimulus generation and result observation on to the IC. The output of the BIST hardware is a simple pass/fail signal. The primary cost associated with BIST is in the area and performance impact of the additional circuitry and the corresponding yield loss. To limit overhead, existing hardware is reconfigured during circuit test. BIST can also be used to reduce system test costs by combining BIST schemes for individual circuits to form a system-level BIST scheme. This can be used not only during system manufacture, but in the field as well. Given the potential advantages, BIST techniques for AMS ICs are naturally of interest. In contrast to digital circuits, the difficulty in AMS IC test is that of signal measurement. That is, precisely applying accurate analog inputs to internal nodes, and observing signal values at internal nodes. BIST methods can potentially resolve the problem of testing AMS components that are not readily accessible from system inputs and outputs. Analog BIST methods have addressed this issue in two ways. Some BIST schemes are direct extensions of DFT schemes that reduce the need for complex test inputs. Simple signals, that need not be accurately calibrated, such as DC inputs and ramps may be sufficient to excite faults and propagate them to observe points. Such signals can be inexpensively generated on-chip. Similarly, checkers in code-based DFT schemes produce simple outputs. These features can be combined to form a low-cost BIST scheme. Functional specification-based test methods demand the generation of more complex inputs, and produce greater amounts of output data. For example, spectral analysis requires the generation and application of accurately calibrated sine waves at a variety of frequencies. Professor Gordon Roberts, a strong proponent of spectral BIST, discusses recent research in this area. This includes the use of programmable sine wave generators to generate test stimuli. On-chip data converters, which are already present, are used to observe analog signals. The methods discussed are applied to circuits of practical importance. As with digital circuits, the primary obstacle to BIST is overhead from the additional circuitry.

Test-Bus Standard   As has happened within ICs, component density at the board level has also increased sharply. The increase in density has made it more difficult to access component inputs and outputs, and to test board-interconnect for faults. Access to components is further reduced by modern packaging techniques such as multichip modules, and flip-chip packages. The conventional bed-of-nails approach to testing is not tenable for many boards. Boundary scan for component input/output pins has been the solution for digital boards. Each component is designed such that its pins can be configured as a shift register in the test mode. At the board level, the scan chains of the components are linked serially. The board has a scan input and output to access the resulting chain. Boundary scan enables the test of nets on the board, as well as the application of functional test vectors to components. The primary costs of using boundary scan are on-chip area and pins devoted to scan. The IEEE standardized one boundary scan methodology. Since a typical board contains components from many vendors, a standard permits these ICs to work together during test.

Given the standard's success, there was great interest in developing a similar standard for AMS ICs. Clearly, chaining analog signals is not an acceptable solution. Thus, the main advantage of a boundary scan scheme is to provide additional access for control and observation of analog signals. This limits the number of signals that can be observed simultaneously. Fortunately, since the number of analog signals is not very large, this is not a significant constraint. Several boundary scan schemes for analog pins were proposed. A single standard was recently proposed by an IEEE working group and Draft 18 passed the first ballot. The standard is based on a hierarchical structure of analog buses that are added to the IC design, and at the board level. The bus structure proposed by the standard can be used within an IC as well. Mr. Stephen Sunter, the vice-chair of the working group, discusses the new standard in detail. More importantly, he discusses how the standard can be used in practice to measure internal signals. The structure of the standard and its application is illustrated using many examples.

Switched Current Circuits Typically, information in circuits is transported through voltages. For example, in digital circuits, a high voltage represents a logic one, and a low voltage represents a logic zero. Switched current circuits are analog circuits in which information is primarily computed and transported through currents. Switched current circuits offer performance and power dissipation advantages over traditional analog circuits. These circuits can also be fabricated using the processes used to manufacture digital ICs. Hence, they may be ideal for integrated data conversion and other high-performance applications. Professor Chin-Long Wey discusses test techniques for switched current circuits.

1.5 Conclusion

Advances in technology and the concurrent economic trends, have spurred a revival of interest in analog and mixed-signal circuit test. Modern research in AMS test differs from earlier efforts in several ways. Integration has strongly influenced the goals and methods of the test process. Researchers have attempted to port methods and algorithms developed for digital integrated circuits to AMS circuits. For example, design-for-test and built-in self-test have already received a lot of attention. Equally significantly, the IEEE is on the verge of releasing an analog extension to the digital boundary scan standard. Many digital test techniques cannot be ported easily because of the significant differences between AMS and digital circuits.

AMS IC test is very difficult for a number of reasons. There is substantial disagreement about the paradigms to be used to test AMS ICs. Designs have traditionally been validated and tested by verifying if they meet functional specifications. Researchers have attempted to adapt specification-based test to integrated circuits. Test stimuli based on such processes are easy to generate, but hard to apply and are very long. Model and structure-based test methods have been very successful with digital IC test. Such approaches enable an algorithmic approach to testing. Correspondingly, there has been significant interest in replicating success by developing a similar paradigm for AMS ICs. This paradigm offers several advantages such as a quantifiable fault coverage, a compact set of test stimuli and so on. The primary obstacle to widespread acceptance is an inability to prove that detecting all modeled faults would imply a circuit that meets all its specifications. The two contrasting paradigms influence all aspects of testing. Efforts in both approaches to testing analog and mixed-signal circuits are reviewed in this book.

References

1.   M. Ismail and T. Fiez, eds., *Analog VLSI: Signal and Information Processing*, McGraw-Hill, 1994.
2.   R.-W. Liu, *Testing and Diagnosis of Analog Circuits and Systems*, Van Nostrand Reinhold, 1991.
3.   R. Gregorian and G. Temes, *Analog MOS Integrated Circuits for Signal Processing*, Wiley Series on Filters: Design, Manufacturing, and Application, New York: Wiley and Sons, 1986.
4.   M. Abramovici, M. Breuer, and A. Friedman, *Digital Systems Testing and Testable Design*, Englewood Cliffs, NJ: Computer Science Press, 1990.

Defect-Oriented Testing

Manoj Sachdev

Manufacturing process defects are one of the major causes of yield loss in ICs. This chapter surveys the advances in the field of defect-oriented analog and mixed-signal testing and summarizes strengths and weaknesses of the method for analog and mixed-signal circuits.

2.1 Introduction

Analog testing poses challenges still to be surmounted by researchers. Several reasons are attributed to the inherent analog test complexity [1,12,18,19] and a number of solutions have been suggested [1-19]. However, in spite of these attempts and proposed solutions, almost all analog circuits are presently tested in a functional manner. The analog test complexity is different from that of digital circuits. The emergence of mixed-signal ICs is further complicating the test issues of analog and mixed-signal circuits.

For any test strategy to succeed in terms of test quality and global applicability, it should have a sound basis. For example, poor performance of the stuck-at model-based digital test schemes amply demonstrate how without a firm basis test, strategies can fail to deliver quality products [30,40]. Defect-oriented testing is proposed as an alternative for quality improvement. In this chapter, we review the work done on defect-oriented mixed-signal testing with the following objectives.

- To propose an analog test methodology based on a firm foundation. The proposed test strategy is based on manufacturing process defects which provide an objective basis for analog fault model development and test generation.

• To assess the effectiveness of the methodology from two standpoints: (a) contribution of defect-oriented test methodology toward testing silicon devices in the production environment and, (b) its contribution towards robust analog design against process defects, quantifying the fault coverage of analog tests, and examining the practicality of analog DFT schemes.

Analog circuits, due to their non-binary circuit operation are influenced by defects in a different manner compared to digital circuits. This poses additional challenges for modeling of faults in analog circuits. In fact, the analog fault modeling is identified as a critical factor in the success of any analog DFT scheme [25]. We also explore the concept of structural test vectors in the analog domain and examine the potential of simple test stimuli in fault detection.

2.2 Previous Work

Analog fault modeling and diagnosis received a lot of theoretical treatment in the late 1970s and 1980s. Duhamel and Rault presented an excellent review of the topic [1]. These theoretical works relied on the characteristic matrix of the circuit under test for testability and diagnosability. Though these methods have a broad scope, their application to specific circuits had not been successful. The analog fault detection and classification can be divided broadly into the following three categories.

2.2.1   Estimation Method

The method can further be subdivided into an *analytical* (or deterministic) method and a *probabilistic* method. In the former, the actual values of the parameters of the device are determined analytically or based on the estimation criteria (physical or mathematical). The least square criterion approach represents this class. Typically, in this approach, a factor of merit, S_i, is associated with each parameter as:

$$S_i = \sum_{j=1}^{m} \{g_j - y_j(X_i)\}^2 \tag{2.1}$$

where g_j is the measured value of the characteristic y_j, and X_i is the vector $x_1, \ldots, x_n$ where the parameters have their nominal values, except for x_i. The factor of merit associated with x_i is taken as the minimum value of S_i. The most likely faulty parameter is the one which, when all other ones are at their nominal value, minimizes the difference between nominal and measured characteristics.

In probabilistic methods the values are inferred from the tolerance of the parameters. For example, the inverse probability method is a representative of this class. Elias [2] applied statistical simulation techniques to select parameters to be tested. On this basis, he also formulated the test limits.

2.2.2 Topological Method

This method is also known as *simulation-after-test* (SAT) method. The topology of the circuit is known and SAT method essentially reverse engineers a circuit to determine the values of the circuit component parameters. A set of voltage and/or measurements are taken and then applying numerical analyses to determine the parameter values [3, 4, 5, 6, 9, 10, 11, 12]. SAT methods are very efficient for soft-faults diagnosis because they are based on a linearized network model. However, this method is computer intensive and for large circuits, the algorithms can be inefficient.

One of the first theoretical studies of the analog circuit fault-analysis problem was initiated by Berkowitz [3]. He mathematically defined the concept of network-element-value solvability and studied the measurement conditions required to solve the problem. Trick et al. [4] and Navid and Willson Jr. [5] proposed necessary and sufficient conditions for the network solvability problem. Trick et al. [4] used only voltage and single frequency sinusoidal input to determine the parameter value for linear passive circuits. Navid et al. suggested that for small signal analysis, non-linear active elements, like transistors and diodes, can be linearized around their operating point. They proposed an algorithm covering the solvability problem for linear passive network as well as active devices. Rapisarda and Decarlo [9] proposed the *tableau approach* for analog fault diagnosis instead of transfer function oriented algorithms. They argued that *tableau approach* with *multi-frequency excitation* would provide simpler diagnostic solution. Salama et al. [10] proposed that large analog circuits can be broken into smaller uncoupled networks by nodal decomposition method. These subnetworks can be tested independently or in parallel. Every subnetwork is associated with a *logical variable* σ, which takes the value of 1 if the subnetwork is good and 0 if it is faulty. Furthermore, every test is associated with a *logical test function* (LTF) which is equal to the complete product of variables σ_{ji}. If the test, T, of the network is a pass, then

$$T_{J_t} \equiv \sigma_{j1} \cap \sigma_{j2} \ldots \cap \sigma_{jk} \qquad (2.2)$$

where

$$J_t \equiv j_1, j_2, \ldots, j_k \qquad (2.3)$$

j_i refers to network S_{j_i}, k is the number of sub-networks involved in the test.

Hemink et al. [11] postulated that the solvability of the matrix depends on the determination accuracies of the parameter. The set of equations describing the relations between parameters and measurements can be *ill-conditioned* due to "almost" inseparable parameters. Further, they contended, that solving such a set of equations inevitably leads to large faults. They overcome this problem by an improved algorithm which finds the sets of separable high-level parameters and computes the determination accuracy of the parameters. Recently, Walker et al. [12] developed a two-stage SAT fault-diagnosis technique based on bias modulation. The first stage, which diagnoses and isolates faulty network nodes, resembles the node fault location method. The second stage, a sub-network branch diagnosis, extracts faulty network parameters. The branch diagnosis is achieved by *element modulation*, a technique to vary the value of the element externally as a modulated element. The diagnostic technique requires a single test frequency and the ability to control the network bias from an external source.

2.2.3   Taxonomical Method

This method is based upon a fault dictionary. This is also known as the simulation-before-test (SBT) method [1, 13, 14, 15, 16, 17, 18, 19]. The fault dictionary is a collection of potential faulty and the fault-free responses. During the actual testing the measured value is compared with the stored responses in the dictionary. A fault is detected if, at least for a set of measurements, the actual response differs from the fault-free response by predetermined criteria. The accuracy of the method depends on the accuracy of the fault dictionary [1].

The fault-free and faulty circuit under test (CUT) responses are measured at certain key points. The number of test points depends on the diagnosis resolution and test stimuli. Schemes based on this method can be segregated according to the input stimuli and fault dictionary construction. For example, this method can be implemented with DC signals [13,15], with various time-domain signals [16] or with AC signals [17]. Milor and Vishvanathan [15] argued that most faults are catastrophic in nature. Thus, DC tests with few test points have potential in detecting such failures. They developed an algorithm to detect catastrophic faults at wafer test. Their algorithm contained two steps:

- Determination of a tolerance box containing process parameters. These parameters are typically specified for a given process (e.g., gate oxide thickness, substrate doping, transistor threshold voltage, etc.).
- Mapping the tolerance box by the sensitivity matrix of the good circuit onto the measurement space such that the good signature of the circuit under test is determined.

For parametric faults, Milor and Sangiovanni-Vincentelli [23] devised an algorithm that selects a subset of circuit specifications for a desired parametric fault coverage. The DC fault dictionary approach is simple but it can not detect purely capacitive or inductive defects. Such defects often give rise to parametric or soft faults which can be more readily detected by a transient or AC dictionary approach. Slamani et al. [18] made a combined dictionary for DC, transient, and AC input stimuli to predict the defective component. They claimed that this method could detect wide-ranging defects, from tolerance deviation to catastrophic faults. Sachdev [20] attempted to make a similar fault dictionary with catastrophic processing defect information using the Inductive Fault Analysis (IFA) [33].

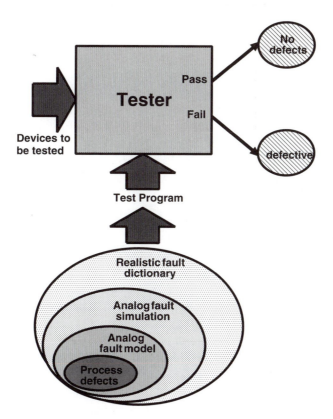

Figure 2-1 *Realistic defect-based test methodology for analog circuits.(Figure courtesy of WKAP, © 1995 [45].)*

2.3 Defect-Based Realistic Fault Dictionary

Application of a defect-oriented approach [33] in solving analog test problems has gained popularity in recent past [20, 24, 25, 26, 44, 45]. It is proposed as one of the alternatives to analog functional testing. However, this attention is not without controversy. What makes this topic so controversial? The critics of IFA-based analog testing are quick to remind that the test issues of analog circuits are more qualitative than quantitative. It is not uncommon to come across an analog circuit having Signal-to-Noise Ratio (SNR) of 100 dB or operation frequency of a few hundred MHz or input-offset voltage less than 20 mV. Secondly, analog circuits often exploit a number of circuit and device level parameters (e.g., transistor matching, layout considerations, transistor sizing etc.) to achieve maximum possible performance under given circumstances. Unfortunately, such clever techniques render the circuit vulnerable to several factors since the maximum possible circuit performance is achievable only under the optimal fabrication and operating conditions. Thirdly, in the case of analog circuits, the range of optimal conditions is substantially narrower than that of their digital counterparts. For example, in digital circuits, typically the critical path is the most sensitive for performance (parametric) degradation. While in analog circuits, the parametric requirement is much higher and widely distributed over the circuit layout. Therefore, any sub-optimal performance of one or more parameters may have a significant impact on the performance. A good test program should test for all such sub-optimal performances. Finally, how comprehensive and accurate is the yield-loss model based upon defects alone in the case of analog circuits? Since these are formidable concerns, according to critics, the functional (specification) testing is the only alternative to ensure the circuit performance, specifications and quality.

On the other hand, those who have faith in IFA-based analog testing will argue that IFA-based testing combines the circuit topology and process defect data to arrive on the realistic fault data specific to the circuit. This information can be exploited by test professionals to generate effective and economic tests. The same information can be used by analog circuit designers to design robust and defect-tolerant circuits and to employ practical, effective DFT schemes. Secondly, this is a structured and globally applicable test methodology which reduces the test generation cost substantially. Finally, they will cite numerous examples of digital domain where IFA-based tests contributed significantly in test simplification and test quality improvement [33]. We address the assessment of analog IFA from two standpoints: (1) contribution of IFA towards testing silicon devices in the production environment, and (2) contribution of IFA toward robust analog design against process defects, quantifying the fault coverage of analog tests, and examining the practicality of analog DFT schemes.

In the classical sense, the defect-based fault dictionary can be categorized as an SBT approach. All fault simulations are carried out before the test. Figure 2-1 illustrates basic concepts of the defect-based fault dictionary. The manufacturing process defects, catastrophic as well as non-catastrophic, form the core of the methodology. Realistic defects are sprinkled over the circuit to determine the realistic fault classes. These faults are simulated with given test vectors. The fault simulations are carried out in a Spice-like simulator to achieve accurate results. Alternatively, if fault simulations at circuit level are not possible owing to the circuit complexity, a high-level model of the circuit may be used. The responses of the fault simulation are compiled into a fault dictionary. A fault is considered detectable if the faulty response differs from the nominal response by a predetermined criterion. Now tests are generated and test programs are made taking the fault dictionary into account. The effectiveness of the fault dictionary depends on several factors. Foremost among them is the defect population and relative probabilities of occurrence of various defects. It is not possible to carry out exhaustive fault simulations with all permutations of defect (fault) impedances. Therefore, the effectiveness of the dictionary depends on how representative are fault models of actual defects and how accurate is the simulator. Finally, the dictionary effectiveness also depends on pass/fail criterion. Nevertheless, the defect-based fault dictionary forms the basis for structured analog testing.

Mathematically, we can define this concept as follows: let $\mathcal{F}$ be the fault matrix of all the faults in a given CUT and let F_0, the first element of the matrix, be the fault-free element. Moreover, let $\mathcal{S}$ be the matrix of stimuli applied at CUT inputs and let $\mathcal{D}$ be the matrix of the fault-free and faulty responses (i.e., the fault dictionary). Furthermore, let us assume that in a given circuit, there are n faults, then the complexity of the fault matrix taking into account the fault-free element as well, is $(n+1)$ by 1. The fault matrix, $\mathcal{F}$ can be written as follows:

$$\mathcal{F} = \begin{bmatrix} F_0 \\ F_1 \\ F_2 \\ \circ \\ \circ \\ F_n \end{bmatrix} \qquad (2.4)$$

For the formulation of the stimuli matrix, let us assume that CUT has m inputs. Therefore, any arbitrary test vector S_i consists of $s_{i_1}, s_{i_2}, ..., s_{i_m}$. In order to simplify the analysis we assume that for any S_i all constituents put together excite the faulty CUT in a particular way. Therefore, the constituents of S_i can be treated as *scalar*. This is not an unreasonable assumption since in analog circuits, unlike digital circuits, the circuit functionality depends on the continuous operation of its components and stimuli. The analysis holds true even in the absence of this assumption: however, it requires rigorous mathematics. Furthermore, in spite of this assumption, one has total freedom to select the constituents of a given (S_i) stimuli. Hence, the stimuli matrix can be formulated as:

$$\mathcal{S} = \begin{bmatrix} S_1 & S_2 & ... & S_i & ... & S_t \end{bmatrix} \qquad (2.5)$$

where t is total number of inputs. The fault dictionary $\mathcal{D}$ is a function of the fault matrix as well as the stimuli matrix.

$$\mathcal{D} = f(\mathcal{F} \times \mathcal{S}) \tag{2.6}$$

For each fault detection mechanism such as voltages on different outputs or dynamic current, formulation of different matrices will be required. Alternatively, like the stimuli matrix, different detection mechanisms can be treated as scalar field of each d_{ij}. Elements of the matrix $\mathcal{D}$ are given as follows:

$$d_{ij} = f_{ij}(F_i \times S_j) \tag{2.7}$$

where $0 \leq i \leq n$ and $1 \leq j \leq t$

At this moment, we simulate the CUT to find out all d_{ij} of the fault dictionary. However, it is possible to compute these elements if function f_{ij} is known. Furthermore, the first row of the $\mathcal{D}$ gives the fault-free responses and the complexity of $\mathcal{D}$ is $(n+1)$ by t.

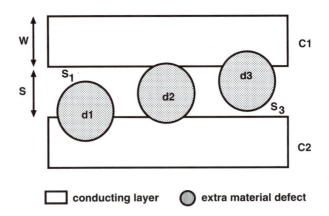

conducting layer   ⬤ extra material defect

Figure 2-2   *Catastrophic and non-catastrophic faults. (Figure courtesy of WKAP, © 1995 [45].)*

2.3.1   Implementation

The implementation issues of the fault dictionary may be segregated into issues (1) related to defects and fault modeling, and (2) related to the analysis flow. The former is concerned with the collection of defect data for a given fab, modeling of defects for a given fault simulator, and so on. The latter is concerned with establishing an analysis flow, determination of pass/fail criterion, *et cetera*.

(a) Process defects and modeling: Defects and their impact on the device performance have been studied in great detail in the literature [27, 28, 29, 30, 31, 32].

Broadly speaking, causes of IC functional failures can be segregated into global and local process disturbances. Global disturbances are primarily caused by defects generated during the manufacturing process. The impact of these global (or manufacturing process-related) defects covers a relatively wider area. Hence, they are detected before functional (or structural) testing by using simple test-structure measurements or supply-current tests. A vast majority of faults that have to be detected during functional (or structural) testing are caused by local defects, or more popularly known as *spot defects* [27]. Since the global defects are relatively easy to detect and are detected by other measurements, we used spot defects for fault-modeling purposes. In a typical single-poly double-metal CMOS process, following are commonly found spot defects:

• Short between wires
• Open in wires
• Pin holes; oxide, gate oxide, *pn*-junction
• Extra contacts or vias
• Missing contacts or vias

Furthermore, spot defects can be categorized into two classes, (1) causing a complete short or open in the circuit connectivity. These are often referred to as catastrophic defects, (2) causing an incomplete short or open in the circuit connectivity. These are often called non-catastrophic or soft defects. Figure 2-2 shows catastrophic and non-catastrophic defects caused by spot defects between two conducting layers C_1 and C_2. Defect d_2 causes a catastrophic short (low-resistance bridge) between both conductors. Therefore, the defect modifies the circuit characteristics and performance drastically. However, defects d_1 and d_3 do not cause complete shorts but reduce the spacing to s_1 and s_3, respectively. Reduced spacing causes high-impedance bridging defects between conductors which can be modeled as a parallel combination of resistance R and capacitance C. The values of R and C are given by the following equations:

$$R = \frac{\rho_{SiO_2} \times S_1}{A} \tag{2.8}$$

$$C = \frac{\varepsilon_{SiO_2} \times A}{S_1} \tag{2.9}$$

In these equations, ρ_{SiO_2} is the resistivity and ε_{SiO_2} is the permittivity of the insulator between the conductors C_1 and C_2. The S_1 is the reduced spacing between the conductors which were otherwise a distance S apart. The area between the defect and conductor is represented by A. The resistance of the short is directly and the capacitance of the short is inversely proportional to distance S_1. Eqn. 2.10 shows the resultant impedance of such a short.

$$Z_{short} = \frac{R}{1 + j2\pi \times fRC} \approx R \tag{2.10}$$

As can be concluded from Eqn. 2.10 that the impedance of the short is not only a function of the spacing S_1 but also depends inversely on frequency and phase relationship of the two conductors. Moreover, at low frequencies, the model of such defects is mainly resistive but at a certain frequency, f_T, their model can become primarily reactive. Thus, the fault model of such defects undergoes a transition and the transition frequency, f_T, depends on the defect geometry, spacing s, the resistivity, and the permittivity of the insulating layer. Therefore, such defects influence the circuit connectivity in different manners. For example, a particular soft defect may have very little impact on low frequencies but at high frequency it may be significant. Figure 2-3 shows a photograph of a non-catastrophic short in a metallization layer. The extra material defect reduces the distance between two metal conductors giving rise to a high-impedance bridging defect between them. However, for most of the applications and technologies the impedance of the short can be approximated as purely resistive.

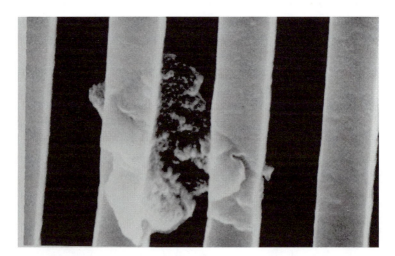

Figure 2-3 *Photograph showing a high-resistive short in a metallization layer. (Figure courtesy of WKAP, © 1996 [56].)*

(b) Fault simulation environment: The block diagram of the environment is shown in Figure 2-4. The process technology data, defect statistics, and the layout of the circuit under investigation form inputs of the environment. The defect-statistics block contains the process defect-density distributions. For example, probability of shorts in metallization is significantly higher compared to that of open defects in diffusion. A catastrophic defect simulator, VLASIC [34] is utilized to determine the realistic fault classes specific to the circuit and layout. VLASIC mimics the sprinkling of defects into the layout in a manner similar to mature, well-controlled production environment. The output of the simulator is a catastrophic defect list.

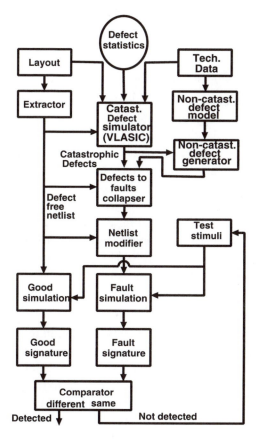

Figure 2-4 *Block diagram of the realistic defect-based testability methodology. (Figure courtesy of WKAP, © 1995 [45].)*

However analog circuits due to a number of reasons are susceptible to non-catastrophic or parametric defects also. Such defects are also often called near-miss defects. We assume that such defects can occur at all the places where catastrophic defects are reported by the defect simulator. However a word of caution: pinhole defects are inherently parametric (high impedance) in nature. Therefore, only shorts and opens in various layers are considered for non-catastrophic defects generations. These defects are appended to the catastrophic defect list. The defect list contains many defects which can be collapsed into a unique fault class. This process is carried out to find out likely fault classes in the layout. Subsequently, each of the fault classes is introduced into a defect-free netlist for fault simulation. This defect-free netlist is extracted from the same layout. For the greatest accuracy the fault simulation is based upon a circuit simulator. The response of the fault simulator is called a fault signature. A fault is considered detected if the corresponding fault signature is different from the defect-free (good) signature by a predetermined threshold. If a faulty response does not differ from the good signature by the threshold, the fault is considered not detected by the stimulus and hence another stimulus is tried. This whole process is carried out for all the faults.

There are a couple of things worth highlighting in the above-mentioned analog fault simulation methodology. First of all, unlike digital circuits, analog circuits lack the binary distinction of pass and fail. In fact, the concept of pass and fail in analog circuits is not clear-cut. It depends on several variables including input stimulus, output measurement parameters (o/p voltage, I_{DD} current, etc.), circuit performance specifications, and permitted environmental conditions (e.g., supply voltage, process, temperature etc.). In other words, there is no absolute reference for fault distinction. A reference has to be evolved for a given circuit under given conditions. This generation of a reference is a tedious exercise and it should be created for each set of input stimuli. The impact of faults is measured against these sets of references. Therefore, a reference response or good signature is a multi-dimensional space and the faulty circuit must exhibit a response outside this space to be recognized as faulty at least by one of the test stimuli.

The graph in Figure 2-5 illustrates this concept. In the graph, two axes form the primary output measurement parameters and the third axis forms an environmental condition (e.g., fabrication process spread). A set of graphs can be plotted essentially showing a possible good signature spread. The good signature spread (shaded area) is generated for each of the given test vectors. A fault is considered detected by a given test vector if the faulty output response of the circuit lies outside the good signature space. For example in Figure 2-6(a), the fault F_1 is detected by test vector S_1 with the output voltage measurement. However, it is not detected by the I_{DD} current measurement since the faulty current lies within the good current spread. On the other hand, same fault is detected by test vector S_2 with output voltage as well as I_{DD} current measurements. The information about fault detection is compiled into a fault dictionary D. Figure 2-6(b) shows the fault dictionary. Rows of the fault dictionary show different fault classes (i.e., $F_1...F_n$) and columns show stimuli (i.e., $S_1...S_t$) with voltage and current as subfields.

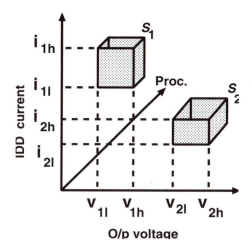

Figure 2-5 *The good signature spread. (Figure courtesy of WKAP, © 1995 [45].)*

Finally, for a structured analog DFT methodology to succeed, effective, efficient test generation is of vital importance. Analog signals are time and amplitude continuous. Therefore, the concept of analog test vectors is not very well defined. For example, in digital domain a binary change in input stimulus is termed as a change in test vector. These vectors are generated in a precise manner covering a predetermined fault set. However, in analog domain, often a test vector is defined as a set of input stimuli required for a particular measurement (specification). The parity between digital and analog test generation can only be restored if the basis of analog test generation is also a predetermined fault set. In this manner, true analog test vectors can be evolved. Furthermore, since all likely fault classes are known, in principle, simple test stimuli can detect the presence (or absence) of a defect.

2.4 A Case Study

A class AB stereo amplifier was selected as the vehicle to see the effectiveness of this methodology. This chip is mass produced for consumer electronics applications. Owing to high volumes and low selling cost, it was desirable to cut down the chip test costs and at the same time maintain quality of the shipped product. It is a three-stage amplifier. The first and second stages are completely differential in nature. The outputs of the second stage feed to the output stage which drives the load of 32 ohms. It was designed in a standard 1.0 micron single-poly double-metal CMOS process. The chip contains two identical amplifiers (channels A and B) and a common biasing circuitry for them. Since both the channels of the class AB stereo amplifier are identical, only one amplifier is considered for testability analysis.

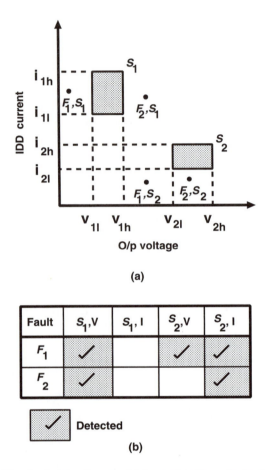

(a)

Fault	S_1,V	S_1, I	S_2,V	S_2, I
F_1	✓		✓	✓
F_2	✓			✓

✓ Detected

(b)

Figure 2-6 *(a) The fault detection; and (b) construction of a fault dictionary.(Figure courtesy of WKAP, © 1995 [45].)*

2.4.1   Fault Matrix Generation

The layout of the amplifier passed through the testability environment. VLA-SIC was utilized to introduce 10,000 defects into the layout. Since most of the defects are too small to cause catastrophic defects, only 493 catastrophic defects resulted in the layout. These defects were further collapsed into 60 unique fault classes. Table 2-1 shows the relevant information about the different fault classes owing to catastrophic defects. A catastrophic short in metal layers was modeled as a resistor with nominal value of 0.2 ohms. Similarly, shorts in poly and diffusion layers were modeled with a resistor of 20 and 60 ohms, respectively. Extra contact and via were modeled with a resistor of 2 ohms. Thick oxide defects were modeled as a resistor of 2K ohms. The gate poly is doped n-type and all the gate oxide shorts occurred in n-channel transistors causing shorts between gates and the sources/ drains of the transistors. Therefore, such shorts were non-rectifying in nature and hence were modeled as a 2K resistor. The n-channel transistor is more susceptible to gate oxide shorts and most of the gate oxide shorts are likely to occur between gate and source/drain [40].

Table 2-1 *Catastrophic fault classes and their fault models. (Table courtesy of WKAP, © 1995 [45].)*

Defect	Number	%	Model (Ohm)
Shorts	22	37	0,2,20,60
Extra contact	10	17	2
Oxide pinhole	15	25	2k
Gox. pinhole	7	11	2k
Junc. pinhole	6	10	2k
Open	0	0	--
Total	60	100	

As mentioned before, soft faults were evolved from hard faults data. Soft faults were generated at locations of shorts and opens in interconnect, contacts, and vias. Therefore, 32 soft fault classes (first 2 rows of Table 2-1) were evolved. Rod-riguez [39] reported that the majority of bridging defects are below 500 ohms. Therefore, the resistance of non-catastrophic defects was chosen as 500 ohms. The capacitance was calculated from the technology data keeping spacing between defect and conductor(s) as 0.1 micron. The computed value is 0.001 pF.

All the catastrophic defects in the defect list generated by VLASIC were shorts in nature caused by extra material, oxide pin holes, or extra contact. None of the defects caused an open circuit. However, given the defect densities for the differ-ent defects in the fabrication process, it was hardly surprising. The shorts in the backend of the process constitute the majority of the spot defects. Furthermore, the layout geometries in analog circuits are often non-minimum size and multiple con-tacts and via contacts are utilized in order to reduce the contact resistance. All this put together made occurrence of an open in the given layout less probable. However in real life, the nature of the above-mentioned defects can vary a great deal and hence no simulation can claim to be exhaustive. Nevertheless, these numbers are consistent with the resistivity of respective layers and published data. Furthermore, for such an analysis, the order of defect resistance is more important than the absolute value.

2.4.2   Stimuli Matrix

For this case study, we divided test signals in three categories: (1) DC stimuli, (2) 1 kHz sinusoid stimuli, and (3) AC stimuli. Often analog circuit functionality depends on continuous operation of all subblocks. Therefore, it is quite likely that a catastrophic fault would change the DC operating point of the circuit and hence will be detected by a DC test stimuli. This may also hold true for some of the high impedance non-catastrophic faults as well. A lower-frequency sinusoid was chosen due to the fact that many fault classes may not be excited by a DC voltage or conditions. Finally, the AC stimuli were chosen because the impact of many non-catastrophic faults is frequency dependent and it is worthwhile to analyze the frequency response as well.

For the simulation of fault classes, the amplifier was put into the configuration shown in Figure 2-7. A load of 32 ohms was placed on the output with a DC blocking capacitor in between. A 2K ohms feedback resistor is placed between output and negative input. Furthermore, full load was put on the output. Before proceeding to fault simulation, the defect-free response was compiled. A fault is considered detected if defect-free and faulty responses differ by at least 1 volt for output voltage measurements and by 0.5 mA for supply current measurements. For AC analysis a fault is considered detected if it modifies the frequency response by 3 dBs. For the DC analysis the positive input was held at 2.5 volts and the negative input was swept from 0 to 5 volts. The output voltage and current drawn from the VDD were sampled when negative input was 1,2,3 volts, respectively. Similarly, for low frequency transient analysis, a 1 kHz sinusoidal signal is applied and root mean square (rms) values of output voltage and IDD current is calculated. For AC analysis different frequency signals on the negative input is applied while the positive input is held at 2.5 volts. In this configuration, the gain of the amplifier is measured.

2.4.3   Simulation Results

Figure 2-8(a) shows the result of fault simulation over catastrophic faults. In this figure obtained results are independently segregated according to the mode of the analysis. On the X-axis, the mode of analysis means the type of input excitation. The detection mechanisms are represented by output voltage, supply current, and gain of the amplifier. The Y-axis shows the percentage of faults detected by each mode of analysis and detection mechanisms independently. For example, DC volt-

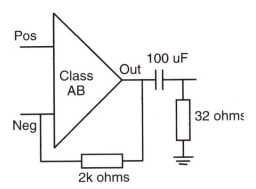

Figure 2-7 *Fault simulation configuration for the class AB amplifier.(Figure courtesy of WKAP, © 1995 [45].)*

age detection of a fault means that the particular fault was detected by output voltage measurement when input excitation was DC. The third column in DC analysis shows how many faults were detected either by DC voltage or by DC current. Therefore, it is the logical OR of first two columns of the same analysis. Similarly, the third column in transient analysis also is the logical OR of first two columns of transient analysis.

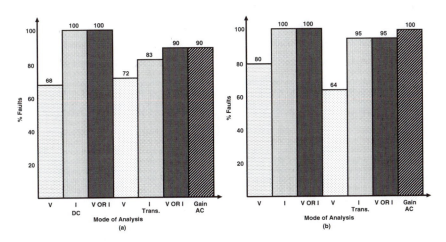

Figure 2-8 *Fault simulation results over: (left) catastrophic defects and (right) non-catastrophic defects. (Figure courtesy of WKAP, © 1995 [45].)*

In DC analysis, 68% of the faults were detected by output voltage. However, the current is a better measure for fault detection and all faults were detected by it. Needless to say that when both detection mechanisms, voltage and current, are considered together, all faults were detected. Though the voltage detection of faults in transient analysis is higher than that of DC analysis, the results, in general, were less encouraging. The 72% faults modified the voltage signature of the device and 83% faults modified the current drawn from the VDD. When both of the mechanisms were considered together, 90% of the faults were detected. Lower than expected performance of transient analysis compared to DC analysis can be attributed to the fact that transient defect-free current or voltage signature is sinusoidal and comparison of two sinusoids on a tester (or in simulations) is more difficult than the comparison of two DC currents. Therefore, in transient analysis, fault detection is carried out manually. If a fault modified the response more than determined threshold, it is considered detected. In AC analysis, 90% of the faults modified the frequency response of the circuit and hence are detected.

Figure 2-8(b) presents the fault simulation results over non-catastrophic faults. The effectiveness of current in DC analysis for faults detection is once again demonstrated. However, for such defects, gain of the amplifier is also an important detection mechanism. All the faults were detected by both analyses. Given the model of these non-catastrophic faults (500 ohms), it was expected.

2.4.4   Silicon Results

Conventionally, devices are tested by verifying a set of DC and AC specifications. The DC specifications include input offset voltage, input bias current, common mode voltage range, output voltage swing, output impedance, output current, and so on. The AC specifications include Total Harmonic Distortion (THD), Signal to Noise Ratio (SNR), slew rate, output power, etc.

A set of 18 passed devices and 497 failed devices with the conventional test process are selected. The passed devices are selected to observe the spread of good signatures compared to simulated thresholds. The comparison is shown in Table 2-2. The values shown outside the brackets represent actual (silicon) and inside the brackets represent simulated thresholds for pass/fail. The actual voltage spread is much smaller than the simulated thresholds, however actual current spread is at least an order of magnitude larger than simulated thresholds. One of the explanations for high current spread is that for this experiment the device is excited in a different manner compared to its typical usage. Therefore, current spread in this configuration was not controlled.

The performance of proposed tests over failed devices is shown in Figure 2-9. Over channel A, DC and transient voltage as well as the gain measurements caught all the faulty devices. The performance of current measurement was less satisfac-

Table 2-2   *Silicon good signature spread. (Table courtesy of WKAP, ©*
1995 [45].)

	DC	Trans.	AC
Voltage	0.15(1) V	0.1(1) V	2(3) dB
Current	9(0.5) mA	5(0.5) mA	—

tory. This difference between simulated and silicon results was due to high current spread in defect-free silicon signature. The DC current measurement was more effective than that of transient. On the other hand in channel B, DC voltage was a poorer method to catch defects compared to DC current measurements. However, transient voltage was more effective than the transient current measurement. The gain measurement over channel B detected 327 faulty devices. In general, channel A was found to be failing more often. No failure analysis was carried out to determine the causes of the differences between both channels. Probably subtle layout differences between both channels are the reason for different failure rates.

Subsequently, the proposed test method was put into the production test environment along with the conventional test method. A test program was evolved in which first, devices were tested with the conventional test method and then with the proposed test method. This exercise was carried out to find out the effectiveness of the method with respect to the conventional test. A total of 1894 devices were tested through the combined test method. The yield of the device was very high and only 11 devices were failed by the conventional test. Out of 11 devices 3 could not be failed by the proposed test method. These devices were again tested with the conventional test method. The table in Figure 2-10 illustrates causes of failures of 3 devices by the conventional test method. As can be seen from the table, all the failures are marginal and very close to the specification limits. The likely origin of such failures lies in inherent process variation and not in spot defects. A test methodology with spot defects as the basis can not ensure detection of such faults. Improved control of the process is one possible solution to reduce such escapes.

The difference in simulation and actual results are quite apparent. Several factors contribute to differences. (1) A finite sample size of defects does not cover the whole spectrum of possible defects; (2) A defect may have many possible impedances; exhaustive simulation of all possible defects and impedances is beyond the capabilities of any state-of-the-art simulator; (3) A circuit simulator has limited

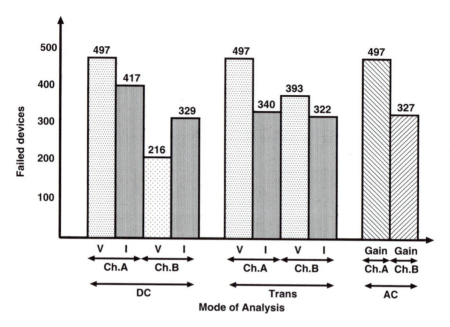

Figure 2-9   *First silicon result for the class AB amplifier. (Figure courtesy of WKAP, © 1995 [45].)*

capability in simulating actual silicon behavior; (4) As mentioned before, higher silicon current spread limits the fault detection capability of the current measurement method. Therefore, DC as well as transient voltage measurements appear to be more effective for fault detection in silicon; (5) Finally, the proposed methodology is based upon spot defects, and global or systematic defects and process variations are not taken into account. Therefore, it is possible that such non-modeled faults are also responsible for differences between simulation and silicon results.

From the second silicon experiment [21, 22] two broad conclusions were drawn: (1) simple tests can detect catastrophic failures; however, detection of some subtle failures is uncertain, and (2) the number of failed devices is not sufficient to draw any meaningful conclusions over the method's applicability in catching real-life faults. More test data, especially on faulty devices, is needed to substantiate claims of IFA-based tests. Therefore, we conducted a relatively large experiment over the same Class AB amplifier devices with the objective to find the effectiveness of the test method on catching real-life failures.

Figure 2-11 illustrates the conducted experiment. A total 3,270 rejected samples of the Class AB amplifier were gathered from a total of 106,784 tested devices by the conventional test method. Only failed devices (3,270) were considered for

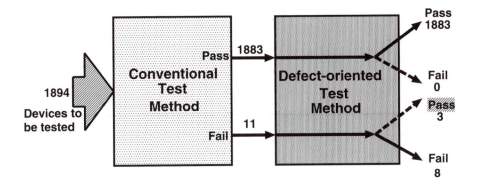

	measured offset	test limit	measuredS/N ratio	test limit
# 1	91.4 mV	<90.0 mV	100.2 dB	>101 dB
# 2	94.5 mV	--	100.1 dB	--
# 3	96.4 mV	--	99.7 dB	--

Figure 2-10   *The results of the second experiment over the class AB amplifier. (Figure courtesy of the IEEE, © 1995 [48].)*

further testing. These devices were tested with the IFA-based test method. Out of this lot, 433 devices passed the test. These passed devices from the IFA-based test method were once again tested with conventional test methods. Results of this test were following: (1) 51 devices passed, the test, and (2) the rest of the devices (433-51=382, 0.4% of total tested devices) failed the test again. These failed devices (382) were subjected to a detailed analysis.

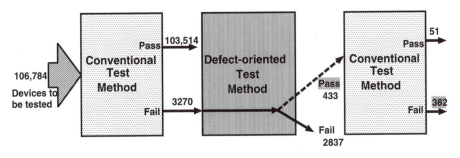

Figure 2-11   *The results of the third experiment over the class AB amplifier. (Figure courtesy of the IEEE, © 1995 [48].)*

2.4.5   Observations and Analysis

Table 2-3 shows the break up of the analysis done over 382 failed devices. The input-offset voltage specification contributes to the maximum number of failures (182, 47.6%) which could not be caught by the IFA-based test method. The Total Harmonic Distortion (THD) specification contributed to the second biggest segment of undetected failures (123, 32.2%). Similarly, SNR measurement failed 20 devices (5.2%). These three categories of failures contribute to the bulk (85%) of the failures that could not be detected by the IFA-based test method. These failures can be attributed to unmodeled faults by the IFA process. For example, any differential amplifier has an inherent offset voltage associated with it which is the source of non-linearity in its operation. Often this offset voltage is minimized by transistor matching, lay-

Table 2-3   Analysis of devices failed in third silicon experiment. (Table courtesy of the IEEE, © 1995 [48].)

Number	Percentage	Failure mechanism
2	0.5	Open/Short
41	10.7	Supply Current
182	47.6	Offset Voltage
1	0.3	Output Voltage swing DC
10	2.6	Common Voltage
0	0.0	Output Voltage AC
20	5.2	S/N ratio
123	32.2	THD at 1 kHz
0	0	X-talk
3	0.8	Ripple rejection
382	100	

out, trimming, and compensation techniques. Besides the process defects, several other factors can manifest themselves as increased offset voltage. Furthermore, increased offset voltage (within specification limits) will give rise to increased non-linearity which will reduce the SNR ratio of the amplifier. This correlation is shown by the table in Figure 2-10. In this table, the device with the highest input-offset voltage shows the lowest SNR, and the device with the lowest input-offset voltage shows the highest SNR. The higher the parametric specifications of an analog circuit, the less effective an IFA-based test is likely to be. This is because natural process variations with higher parametric requirements will contribute to a larger number of device failures. Since these are unmodeled faults, the effectiveness of the IFA-based test is lowered. Furthermore, the effectiveness of a process defect-based yield loss model diminishes significantly with increasing parametric requirements. Therefore, a test based solely on process defects is not sufficient for ensuring the specifications of the device with high parametric specifications.

2.5 IFA-based Fault Grading and DFT for Analog Circuits

One of the major issues faced by analog testing is how to quantify the existing test methods (e.g., functional tests) against the manufacturing process defects. In analog circuits, as we have seen from previous experiments, the functional or specification-based testing can not be eliminated completely in favor of simple DC or AC tests for circuits with high parametric requirements. Furthermore, popularity of mixed-signal devices has compounded the importance of analog test issues. The testing of analog blocks in a sea of digital logic is becoming an increasingly difficult task. Two major issues pose difficulties. Firstly, limited controllability and observability conditions for analog blocks are causing an increase in test complexity and cost. Secondly, in digital domain, a large number of test methods (e.g., functional, structural, I_{DDQ}) and DFT techniques (e.g., scan path, macro test) are available for quantifying and improving the fault coverage. Furthermore, Automatic Test Pattern Generation (ATPG) techniques have reduced the test generation cost for digital circuits significantly. Analog testing lacks such tools and techniques: therefore, analog testing is becoming a bottleneck in the testing of mixed-signal ICs in terms of cost and quality.

The quality of the test, and hence the tested device, depends heavily on the defect (fault) coverage of the test vectors. Therefore, it is of vital importance to quantify the fault coverage. Since the fault coverage of the test vectors on various building blocks of a design is largely unknown, the benefits of any DFT scheme can not be ascertained with confidence. Furthermore, one can not conclude where DFT is needed most. This lack of quantifying information has resulted in abuses of digital DFT schemes in analog domain and is probably one of the important contributing factors in the demise of analog DFT schemes. In this section, we demonstrate how the IFA technique can be exploited to fault grade given (conventional) test vectors. Once, the relative fault coverage of different blocks is known by given test vectors, an appropriate DFT scheme can be applied to the areas where fault coverage of existing test methods is relatively poor. This is demonstrated with an example of a flash A/D converter.

2.5.1   A/D Converter Testing

An A/D converter is normally tested for DC and AC performance. The DC tests typically test for offset voltage and full-scale errors. Static Differential Non-Linearity (DNL) and Integral Non-Linearity (INL) measurements are performed by slowly varying the input signal such that the DC operating point is reached for each measurement. On the other hand, dynamic tests are performed to test for dynamic range, conversion speed, SNR, dynamic DNL, dynamic INL, Bit Error Rate (BER), and so on. These dynamic specifications are often tested by performing BER measurement, Code Density Measurement (CDM), Beat frequency measurement, SNR, or THD measurement. Performance specifications for data converters are discussed in more detail in Chapter 5.

The code density measurement is an effective way of testing A/D converters. The static DNL and INL of the converter can be computed from this measurement. At the input of an A/D converter a waveform is applied. The amplitude of this waveform is slightly larger than the full scale value of the converter. As the waveform traverses from zero to full amplitude, different output codes appear at the output of the A/D converter at different time instances. For an accurate measurement, at least 8 to 16 codes per level are needed [49,50]. This is achieved by repeating the test for a number of cycles of the waveform. Often a triangular waveform is applied because with triangular input waveform every code should have equal density. If a larger or smaller number of codes are found in the CDM, it shows the presence of DNL. A fault is considered detected by CDM if it resulted in more or less (pre-specified criterion) occurrences of a given code.

For the SNR, THD, and SINAD measurements, a sine wave is applied at the input of a converter and output codes are measured. SINAD is defined as the signal-to-noise plus distortion ratio. In order to randomly distribute the quantization error over the measurement, the ratio of signal frequency to the sampling frequency is given by Eqn. 2.11.

$$\frac{F_{signal}}{F_{sample}} = \frac{M}{N} \tag{2.11}$$

M and N are mutually prime integers, and N is the number of samples taken. Mahoney [50] called it M/N sampling.

2.5.2   Description of the Experiment

An 8-bit flash A/D converter [23] is utilized for this experiment. However, the 8-bit flash A/D converter is too complex even for fault-free simulation at the Spice level. Nevertheless, for the accuracy of the analysis, the circuit-level simulation was considered as an absolute requirement. Therefore, a 3-bit model of the converter at the Spice level was made. This model has only 8 (instead of 256) comparators which could be simulated in reasonable time.

IFA is performed over all basic cells of the converter to find out the fault classes. The 3-bit model has eight comparators, one clock generator, one biasing generator, and one reference ladder. For the fault simulation purposes, a fault is introduced in one of the instances of a basic cell in the model. This is shown in Figure 2-12. For example, faults were introduced one by one in only one out of eight comparators at a time. This was under the single fault assumption in the design. The complete A/D model is fault simulated for all comparator faults. Once, the complete fault list for the comparator was fault-simulated, faults were introduced in another cell. In this way, all the likely faults in the circuit were fault-simulated.

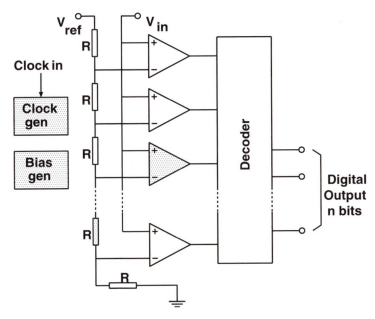

Figure 2-12   *Block diagram of an A/D converter. (Figure courtesy of the IEEE, ©*
1995 [48].)

2.5.3   Fault Simulation Issues

In analog circuits, fault simulation is a laborious exercise. For each fault, a separate simulation is to be run. A relatively higher degree of human interaction, interpretation is required for analog fault simulation. Furthermore, there are analog fault simulation issues which should be carefully addressed.

The simulation environment is considerably slower from that of a tester. A test that takes a fraction of a milli-second on a tester to perform may cost several minutes in simulation environment at Spice level. Furthermore, since fault simulation is to be performed over the complete fault set, the total time for the analysis may become prohibitively large. Due to time constraints, for the CDM, we reduce the average number of codes to 5 and apply a slow ramp such that on every 5th clock cycle a new output code is generated. In other words, for 5 clock cycles, the fault-free converter is supposed to have the same output code. Even then, single fault simulation over the 3-bit A/D model took 8 CPU (HP 700) minutes with a CDM test. We considered a fault detected by CDM if it resulted in more than 7 or less than 3 occurrences of a given code. The SNR, THD, and SINAD measurement takes 45 to 50 CPU minutes for single fault simulation. We selected SINAD instead of SNR as fault detection criterion. The fault simulation using BER could not be performed since it was taking even more time than the SINAD test. The DNL and INL measurement were carried out using the data of SINAD tests.

Secondly, we utilized a verification test system, MixTest [51], to compute DNL, INL, and SINAD fault simulations. A fault is detected by DNL, INL, or SINAD if the computed value differs from the golden device simulation with a predetermined threshold. The sampling frequency of the converter was 20 MHz. To randomly distribute the quantization errors, a fraction of the sampling frequency {(31/128)*20 MHz} was selected as the input frequency.

Thirdly, the setting up of the thresholds for fault detection in simulation environment has to be done carefully. For example, the criterion of 1 LSB for DNL and INL measurements is no longer valid for a 3-bit model of an 8-bit converter. For the original converter, 1 LSB is (2V/256) 7.8 mV. For a 3-bit model, 1 LSB amounts to (2V/8) 250 mV which is substantially more than 7.8 mV. Owing to the constraints of simulation environment, we selected 0.1 LSB (25 mV) as the detection threshold. Though this may be a little conservative but we assumed that if a fault is detected against relaxed threshold criterion, it will certainly be detected in the production environment against much tighter limits. Similarly, for a 3-bit converter, theoretical SNR should be 19.82 dBs and the theoretical SINAD should be somewhat lower than this value. The SINAD for the 3-bit model in fault-free simulation was found to be 18.05 dBs. Once again, we took fairly conservative values for the fault detection. A fault is considered detected by SINAD measurement if the SINAD of the converter was less than 17.5 dBs.

2.5.4   Fault Simulation Results

Table 2-4 compiles the results of the fault simulation over a 3-bit A/D converter model. In the comparator, 157 fault classes were simulated with the CDM test. A set of 112 faults were detected by the CDM test. Because of the large fault simulation times, only those faults which were not detected by CDM (157-112=45) were fault-simulated with the SINAD, DNL, and INL tests. This is further justified by the fact that the CDM test is a simplified version of the SINAD, DNL and INL tests. Therefore, if a fault is detected by CDM, it is likely to be detected by these tests. The DNL test was the most effective (25/45) test in catching the rest of the undetected faults in the comparator. INL and SINAD, detected 21/45 and 19/45 faults, respectively, in the comparator. Nearly 11% of the faults (17/157) in the comparator were not detected by any of these measurements. In the case of the clock generator, 59 fault classes were simulated. The CDM could detect 32 of the simulated faults. The performances of DNL, INL, and SINAD over the remaining undetected faults (27) were relatively poor compared to the comparator. As a result (11/59) of the clock generator faults remained undetected. The performance of conventional tests were the poorest on the bias generator. A total of 50 fault classes were simulated in the bias generator. The CDM could detect only 16. The performance of

Table 2-4 *Fault simulation results on the flash A/D converter. (Table courtesy of the IEEE, © 1995 [48].)*

Test / Cell	Code density measurement	SINAD	DNL	INL	Undetected faults
Comparator	112 (157)	19 (45)	25 (45)	21 (45)	17
Clock generator	32 (59)	10 (27)	14 (27)	10 (27)	11
Bias generator	16 (50)	8 (34)	14 (34)	9 (34)	18
Ref. ladder	16 (19)	1 (3)	1 (3)	1 (3)	2

DNL was marginally better. It detected 14 out of the remaining (34) undetected faults. On the whole, 36% of the total faults remained undetected. In the reference ladder, 19 fault classes were simulated. The CDM was an effective test and detected 16 fault classes. DNL, INL, and SINAD detected 1 fault class of 3 undetected fault classes and the other 2 remained undetected.

2.5.5   Analysis

Nearly 20% of the faults in the clock generator and 36% of the faults in the bias generator are not detected by the popular A/D specification tests. On the other hand nearly 90% of the faults in the comparator and resistor ladder network are detected by the same tests. The difference in fault coverage is not difficult to understand. Most of the conventional specification (conversion speed, SNR, DNL, INL, BER, CDM, etc.) tests are targeted towards faithfulness of the data path (i.e., analog input, reference ladder, comparator, decoder, digital output). There is hardly any test which explicitly tests for the control path (i.e., the clock and the bias generators). These blocks are assumed to be tested implicitly. Poor controllability and observability are other reasons for undetected faults in these blocks. The outputs of these cells are not directly observable. If the faulty bias or clock generator output causes the comparator to behave in an incorrect manner, then the fault is detected by the tests. However, the faults that modify the behavior of the control path marginally are hard to detect and often require testing the complete dynamic range in input amplitude and frequency domains. Furthermore, comparators are often designed to withstand the parametric variations of the clock and biasing to optimize the yield. Such a behavior in the comparators has a fault-masking impact.

Different specification tests have different fault coverage. The relative fault coverage of different tests are illustrated in Figure 2-13. As explained in the previous sub-section, we considered only faults that were not detected by the CDM test for this analysis. Though most of the faults are detected by all tests (DNL, INL, SINAD), DNL is the most effective test. Some faults are only detected by SINAD. However, INL does not detect any fault that is not detected by the other two tests. Nevertheless, we should keep in mind that INL is not a redundant test. The INL is an effective test to detect parametric variations in the reference ladder: for example, those parametric variations which do not cause appreciable shift in the DNL but affect the whole reference ladder.

2.5.6   DFT Measures

The question is how can the fault coverage be improved without sacrificing the performance of the converter. Measuring the quiescent current (I_{DDQ}) may be one solution. However, in analog circuits, unlike digital circuits, the I_{DDQ} is not in the sub-μA range. Therefore, its detection capability is limited. Alternatively, the A/D converter should be designed such that all high-current dissipating paths either be switched off or bypassed for I_{DDQ} testing. Hence, I_{DDQ} testing can be made an effective test method in fault detection. However, the design of such a converter is a non-trivial design effort.

Alternatively, innovative voltage DFT techniques can be applied which do not cause performance degradation while improving the fault coverage. For example, a

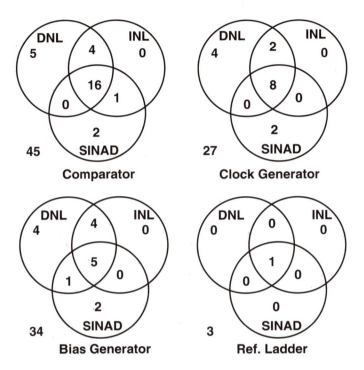

Figure 2-13 *Fault detection capabilities of different tests. (Figure courtesy of the IEEE, © 1995 [48].)*

DFT scheme for the clock driver can be explained with the help of Figures 2-14 and 2-15. Figure 2-14 shows 4 clocks generated from the clock generator. These signals are Boolean in nature but their timing relationship with each other is non-trivial and extremely important for good functioning of the A/D converter. For the DFT solution, we exploit the knowledge of pre-defined timing relationship between the different clocks. Typically, large number of faults degrade the timings of clock signals, if we take a logical AND of these signals (Figure 2-15), we get an output pulse whose position and width is known. Now, most of the faults causing timing and/or stuck-at behavior will be detected by the pulse position and/or width. The number of pulses within the clock cycle and their position from the Clock_in were the fault detection criterion. More than 95% of the faults influenced output(s) of the clock generator and were detected within 2 clock cycles. Therefore, the test method detects faults quickly and provides the diagnostic information. There could be a variety of implementations to extract different attributes of periodic signals. It cost approximately 10 logic gates for the particular implementation. The number of gates is a trade-off between required diagnostics and the cost of implementation. The

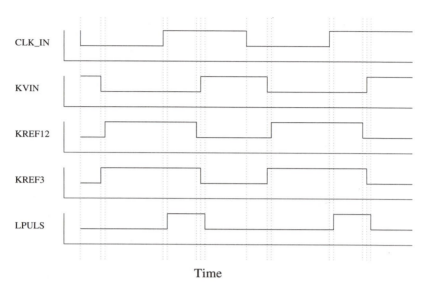

Time

Figure 2-14   *Input clock signal and various generated clocks for the flash converter. Figure courtesy of the IEEE, © 1995 [48].)*

number of gates can be reduced if the critical signal spacing requirements are known from the designer in advance. For example, signals KVIN and KREF12 should be nonoverlapping. Hence, only the critical timings are generated by the Boolean operations.

Similarly, a DFT solution for faults in the bias generator is shown in Figure 2-16. The bias generator provides a set of stable bias signals to the comparator, etc. In the case of the flash A/D converter, 4 bias signals are generated. These are stp1 (3.2V), stp2 (3.2V), stp3 (3.4V), and Vbias (2.3V). Each of the biasing voltages is applied to a p-channel transistor that is individually gated by n-channel transistors. This scheme allows measurement of quiescent current through each or multiple paths. Defect-free quiescent currents through the components can be computed. Now, if the presence of a defect in the biasing network influences the voltage level of any of the biasing signals, it is translated to current which can be measured. Nearly 80% of the bias generator faults could be detected by these simple measurements. Here, it is worth mentioning that the popular and expensive (conventional) test method could detect only 60% of these faults.

2.6 High-Level Analog Fault Models

Defect oriented fault analysis of analog building blocks is a tedious task owing to lack of tools and high computation time. Spice level fault simulation often

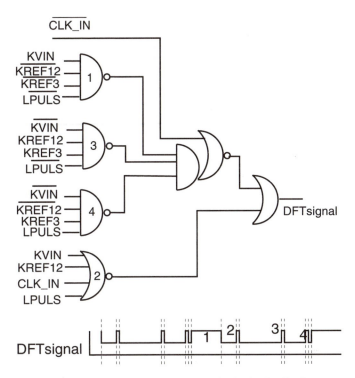

Figure 2-15 *DFT schemes for clock generator faults in the flash converter. (Figure courtesy of the IEEE, © 1995 [48].)*

requires prohibitively large simulation time. Therefore, there have been attempts to create higher-level fault models based upon the realistic defects such that the simulation time may be reduced [42,43,53]. Harvey et al. [42] carried out a defect oriented analysis over a Phase Locked Loop (PLL). Fault-free simulation of PLLs takes enormous CPU time; therefore, for fault simulation a need for higher-level fault models was recognized. The PLL was also divided into several macros (Figure 2-17) for which simulation models at the behavioral level were developed. Although the simulation time per macro was reduced significantly, as can be seen in Table 2-5, it still took several hours of CPU time to simulate the locking behavior, whereas circuit-level simulation was not feasible..

For the IFA analysis, faults were inserted into each macro in turn. To ensure correct fault behavior, the macro being analyzed was replaced by its circuit-level description. Only a few of the faults had to be simulated in a functional way, including all models of the other macros, since most faults caused a hard failure which was already identified in the simulation of the macro being analyzed. The full analysis

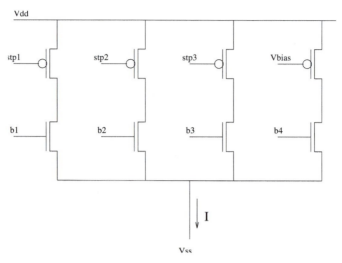

Figure 2-16 *DFT scheme for bias generator faults in the flash converter.(Figure courtesy of the IEEE, © 1995 [48].)*

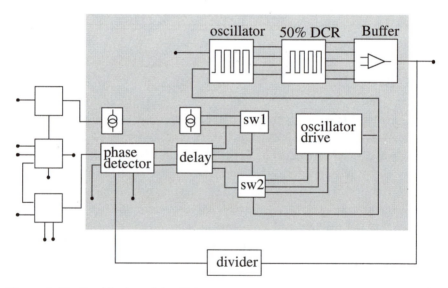

Figure 2-17 *Partitioning of the PLLs into macros. (Figure courtesy of the IEEE, © 1994 [42].)*

revealed that the functional test, comprising locking time and capture-range measurements, will detect about 93% of the faults. Alternative tests were evaluated with respect to customer quality requirements. This revealed that the remaining 7% faults can be detected by using power supply voltage levels outside the specified operational range, which extends from 4.5 to 5.5 V. Application of a supply voltage of 3 Volts changes the circuit sensitivities [54] and thus enables the detection of the other faults [42]. Such tests can only be used when the fault-free response can unambiguously be identified.

In a similar manner, the anti-side tone module of the PACT IC, used for telephone applications (PCA1070 [55]) was subjected to fault analysis. The anti-side tone module consists of two programmable switched capacitor filters within a feedback loop (see Figure 2-18). Being programmable, the IC can be used in various countries with different statutory requirements, but the necessity of the repetition of functional tests for different settings means that a lot of time is required for testing.

Table 2-5 *Comparison of circuit and behavioral level simulation times. (Table courtesy of the IEEE, © 1994 [42].)*

Block	Circuit CPU time (s)	Model CPU time (s)	Speed advantage factor
Oscillator	3082	134	23
50% DCR	803	527	1.5
Buffer	96	4	24
Phase detector	373	14	27
Delay	82	5	16
Current mirror 1	76	3	25
Current mirror 2	106	3	35
Current switch 1	42	3	14
Current switch 2	72	2	36
Oscillator drive	58	6	10

Goal of the analysis of the anti-side tone module was to evaluate the effectiveness of the existing functional tests applied in production testing. Again, the macro-oriented divide and conquer approach was used. For analysis of the ZS-filter, the other macros are modeled at a level optimized in terms of simulation speed. The assembly of all macros together was used in the fault simulation.

The most complex macro is the ZS-filter, a programmable switched capacitor filter, whose programmed digital setting determines the number of switched capacitors that are activated. It takes a very long time to simulate this filter with a circuit simulator due to the combination of the time discrete character of the filter and the functional test signals. Because the simulations have to be repeated for each fault and for various programmed settings, the required simulation time reaches unacceptable levels. By replacing the switched capacitors by equivalent models consisting of resistors and current sources, a time continuous equivalent is obtained, which allows for fast simulations. This equivalent model was used to analyze all faults in the decoding logic, the operational amplifiers, the resistive divider, etc. Faults affecting the clock signals of the switched capacitors and faults inside those capacitors were analyzed separately.

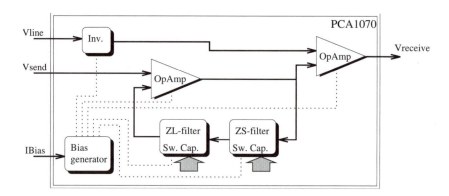

Figure 2-18   *The anti-side tone module of the PCA 1070.*

Analysis of the 55 most likely faults resulted in five faults not detected by the functional tests. Two of these could be detected by adding an additional functional test, but the remaining three faults were inherently undetectable. However, further analysis of these five faults revealed that they can be eliminated by minor changes in the routing of the layout.

2.7 Discussion: Strengths and Weaknesses of IFA-Based Tests

On the basis of the above-mentioned experiments, the following comments regarding strengths and weaknesses of IFA over analog circuits are made. Some of these comments are specific to the Class AB amplifier and others have some general applicability for IFA-based analog testing.

- An IFA-based test method is based upon process defects which contrast with the conventional, specification-based analog test method. The IFA-based test method is structured, therefore, has a potential for quicker test generation. Here it is relevant to mention that the specification-based analog test generation for complex mixed-signal ICs may take many man years of effort. Though, IFA based test generation also requires considerable effort as well as resources, it is faster than the ad hoc test generation.

- The IFA-based tests are simpler and their requirements for test-infrastructure is substantially less compared to the specification-based tests. Therefore, the majority of such tests can be carried out on inexpensive testers. A vast majority of faults is detected by simple, DC, Transient, and AC measurements. For example, presently, the Class AB amplifier devices are tested with a combination of an IFA- based test and a limited functional test. The combined test method results in an estimated saving of 30%.

- The number of escapes (382) of the IFA-based method amounts to 0.358% of tested devices (or 3,580 PPM). Clearly, it is unacceptably high for any device. A limited specification test, as mentioned above, with an IFA-based test may be advantageous in quality improvement while test economics are maintained to an extent.

- The number of escapes can be reduced by a rigorous control of the fabrication process. The basis of IFA is a given set of process defects. However, this basis is not absolutely stable because of the process dynamism. A new defect type may be introduced into the set if the process is unstable or improperly monitored. A better process control (higher Cp and Cpk) will increase the effectiveness of the IFA-based test.

- Effective test generation and limit setting are of crucial importance to the success of IFA-based testing. For example, although supply current measurements have been implemented in IFA tests, a substantial amount of devices (41) passed the test but failed the supply-current test in the conventional test method. This is because the test limits in IFA-based tests are determined more or less arbitrarily. The measured current on 187 good samples suggests that test limits should be more stringent. The same holds true for other detection-thresholds also. Setting of 1 V or 3 dB thresholds for fault detection is not stringent enough to ensure high parametric fault requirements for the amplifier. More research is needed for test pattern generation and threshold settings.

- Design insensitivity against process variations also contributes towards the effectiveness of the IFA-based test vectors. IFA-based tests are ill-suited for design characterization.

- The Class AB amplifier is an audio amplifier with very high parametric requirements and a relatively small number of transistors. IFA-based test

methods are more successful for circuits or ICs where the parametric requirement is relatively low and functional complexity is high. For such complex
ICs, functional testing is not enough and the IFA test may form the main segment of testing. On the other hand, for high-performance analog ICs the IFA-
based simple test may form the basis of wafer sort, therefore, rejecting all
potential defective devices. The subsequent limited functional test will be
applied on only potentially good devices. The combination of these two will
not only improve the economics of the testing but will also result in better
quality of the tested devices.

• Quantifying the fault coverage of a given set of test vectors on an analog circuit is an unexplored area. The IFA-based test generation provides a methodology by which test vectors and designs can be fault-graded. Once, fault
coverage of different tests are known, ordering of tests may improve test economics while tests that do not contribute to fault detection may be discarded.
Furthermore, the impact of test vectors on outgoing quality can also be quantified.

• IFA-based test methods are limited by the availability of CAD software tools
and require high computer resources in terms of CPU power and data storage.
A substantial analysis effort is needed before an IFA-based test method may
emerge from analysis. Furthermore, due to computational and CAD tool-
related constraints only cells and macros can be analyzed. Therefore, ideally
this analysis should be carried out in the design environment on a cell by cell
basis. A bigger design should be partitioned into suitable, smaller, segments
for this analysis.

References

1. P. Duhamel and J.C. Rault, "Automatic test generation techniques for analog circuits and systems: a
    review," *IEEE Transactions on Circuits and Systems*, vol. CAS-26, no. 7, pp. 411-440, July 1979.

2. N.J. Elias, "The application of statistical simulation to automate the analog test development,"
    IEEE Transactions on Circuits and Systems, vol. CAS-26, no. 7, pp. 513-517, July 1979.

3. R.S. Berkowitz, "Conditions for network-element-value solvability," *IRE Transactions on Circuit
    Theory*, vol. CT-9, pp. 24-29, March 1962.

4. T.N. Trick, W. Mayeda, and A.A. Sakla, "Calculation of parameter values from node voltage mea-
    surements," *IEEE Transactions on Circuits and Systems,* vol. CAS-26, no. 7, pp. 466-474, July
    1979.

5. N. Navid and A.N. Wilson, Jr., "A theory and an algorithm for analog fault diagnosis," *IEEE Trans-
    actions on Circuits and Systems*, vol. CAS-26, no. 7, pp. 440-456, July 1979.

6. R.W. Priester and J.B. Clary, "New measures of testability and test complexity for linear analog
    failure analysis," *IEEE Transactions on Circuits and Systems*, vol. CAS-28, no. 11, pp. 1088-1092,
    November 1981.

7. V. Visvanathan and A. Sangiovanni-Vincentelli, "Diagnosability Of Nonlinear Circuits And Sys-
    tems-part I: the DC case," *IEEE Transactions on Circuits and Systems*, vol. CAS-28, no. 11, pp.

1093-1102, November 1981.

8. R. Saeks, A. Sangiovanni-Vincentelli, and V. Vishvanathan, "Diagnosability of nonlinear circuits and systems-part II: dynamical case," *IEEE Transactions on Circuits and Systems*, vol. CAS-28, no. 11, pp. 1103-1108, November 1981.

9. L. Rapisarda and R.A. Decarlo, "Analog multifrequency fault diagnosis," *IEEE Transactions on Circuits and Systems*, vol. CAS-30, no. 4, pp. 223-234, April 1983.

10. A.E. Salama, J.A. Starzyk, and J.W. Bandler, "A unified decomposition approach for fault location in large analog circuits," *IEEE Transactions on Circuits and Systems*, vol. CAS-31, no. 7, pp. 609-622, July 1984.

11. G.J. Hemink, B.W. Meijer, and H.G. Kerkhoff, "TASTE: a tool for analog system testability evaluation," *Proceeding of International Test Conference*, pp. 829-838, 1988.

12. A. Walker, W.E. Alexander, and P. Lala, "Fault diagnosis in analog circuits using elemental modulation," *IEEE Design & Test of Computers*, pp. 19-29, March 1992.

13. W. Hochwald and J.D. Bastian, "A DC approach for analog dictionary determination," *IEEE Transactions on Circuits and Systems*, vol. CAS-26, no. 7, pp. 523-529, July 1979.

14. A.T. Johnson, Jr., "Efficient fault analysis in linear analog circuits," *IEEE Transactions on Circuits and Systems*, vol. CAS-26, no. 7, pp. 475-484, July 1979.

15. Linda Milor, and V. Visvanathan, "Detection of catastrophic faults in analog integrated circuits," IEEE *Transaction on Computer Aided Design of Integrated Circuits and Systems*, Vol. 8, pp. 114-130, February 1989.

16. H. Sriyananda and D.R. Towill, "Fault diagnosis using time-domain measurements," *Radio and Electronic Engineer*, vol. 9, No. 43, pp. 523-533, September 1973.

17. A. Pahwa and R. Rohrer, "Band faults: efficient approximations to fault bands for the simulation before fault diagnosis of linear circuits," *IEEE Transactions on Circuits and Systems*, vol. CAS-29, no. 2, pp. 81-88, February 1982.

18. M. Slamani and B. Kaminska, "Analog circuit fault diagnosis based on sensitivity computation and functional testing," *IEEE Design & Test of Computers*, pp. 30-39, March 1992.

19. B.R. Epstein, M. Czigler, and S.R. Miller, "Fault detection and classification in linear integrated circuits: an application of discrimination analysis and hypothesis testing," *IEEE Transactions on Computer Aided Design of Integrated Circuits and Systems*, vol. 12, no. 1, pp. 102-112, January 1993.

20. M. Sachdev, "Catastrophic defect oriented testability analysis of a class AB amplifier," *Proceedings of Defect and Fault Tolerance in VLSI Systems*, pp. 319-326, October 1993.

21. K.D. Wagner and T.W. Williams, "Design for testability of mixed signal integrated circuits," *Proceeding of International Test Conference*, pp. 823-828, 1988.

22. T.M. Souders and G.N. Stenbakken, "A comprehensive approach for modeling and testing analog and mixed signal devices," *Proceeding of International Test Conference*, pp. 169-176, 1990.

23. Linda Milor and A. Sangiovanni-Vincentelli, "Optimal test set design for analog circuits," *International Conference on Computer Aided Design*, pp. 294-297, 1990.

24. Mani Soma, "Fault modeling and test generation for sample and hold circuits," *International Symposium on Circuits and Systems*, pp. 2072-2075, 1991.

25. Mani Soma, "An experimental approach to analog fault models", *Custom Integrated Circuits Conference*, pp. 13.6.1-13.6.4, 1991.

26. Anne Meixner and W. Maly, "Fault modeling for the testing of mixed integrated circuits", *Proceeding of International Test Conference*, pp. 564-572, 1991.

27. W. Maly, F.J. Ferguson, and J.P. Shen, "Systematic characterization of physical defects for fault analysis of MOS IC cells," *Proceeding of International Test Conference*, pp. 390-399, 1984.

28. W. Maly, W.R. Moore, and A.J. Strojwas, "Yield loss mechanisms and defect tolerance," *SRC-*

CMU Research Center for Computer Aided Design, Dept. of Electrical and Computer Engineering, Carnegie Mellon University, Pittsburgh, PA 15213.

29. W. Maly, A.J. Strojwas, and S.W. Director, "VLSI yield prediction and estimation: a unified framework," *IEEE Transactions on Computer Aided Design*, vol. CAD-5, no.1, pp 114-130, January 1986.

30. W. Maly, "Realistic fault modeling for VLSI testing," *24th ACM/IEEE Design Automation Conference*, pp.173-180, 1987.

31. C.F. Hawkins and J.M. Soden, "Electrical characteristics and testing considerations for gate oxide shorts in CMOS ICs," *Proceeding of International Test Conference*, pp. 544-555, 1985.

32. J.M. Soden and C.F. Hawkins, "Test considerations for gate oxide shorts in CMOS ICs," *IEEE Design & Test of Computers*, vol. 2, pp. 56-64, August 1986.

33. J.P. Shen, W. Maly, and F.J. Ferguson, "Inductive fault analysis of MOS integrated circuits," *IEEE Design and Test of Computers,* vol. 2, no. 6, pp. 13-26, 1985.

34. H. Walker, and S.W. Director, "VLASIC: A catastrophic fault yield simulator integrated circuits," *IEEE Transactions on Computer Aided Design of Integrated Circuits and Systems*, vol. CAD-5(4), pp. 541-556, October 1986.

35. M. Syrzycki, "Modeling of spot defects in MOS transistors," *Proceedings International Test Conference*, pp. 148-157, 1987.

36. E.M.J.G. Bruls, "Reliability aspects of defects analysis," *Proceedings of European Test Conference*, pp. 17-26, 1993.

37. H.H. Huston and C.P. Clarke, "Reliability defect detection and screening during processing - theory and implementation," *Proceedings of International Reliability Physics Symposium*, pp. 268-275, 1992.

38. R.H. Williams, and C.F. Hawkins, "Errors in testing," *Proceedings of International Test Conference*, pp. 1018-1027, 1990.

39. R. Rodriguez-Montanes, E.M.J.G. Bruls, and J. Figueras, "Bridging defects resistance measurements in CMOS process," *Proceeding of International Test Conference*, pp. 892-899, 1992.

40. J.M. Soden, and C.F. Hawkins, "Electrical properties and detection methods for CMOS IC defects," *Proceedings of European Test Conference*, 1989, pp. 159-167.

41. M. Soma, "A design for test methodology for active analog filters," *Proceedings of International Test Conference*, 1990, pp. 183-192.

42. R.J.A. Harvey, A.M.D. Richardson, E.M.J. Bruls, and K. Baker, "Analogue fault simulation based on layout dependent fault models," *Proceedings of International Test Conference*, pp. 641-649, 1994.

43. F.C.M. Kuijstermans, M. Sachdev, and L. Thijssen, "Defect oriented test methodology for complex mixed-signal circuits," *Proceedings of European Design and Test Conference*, pp. 18-23, 1995.

44. M. Sachdev, "Defect oriented analog testing: strengths and weaknesses," *Proceedings of 20th European Solid State Circuits Conference*, pp. 224-227, 1994.

45. M. Sachdev, "A defect oriented testability methodology for analog circuits," *Journal of Electronic Testing: Theory and Applications*, vol. 6, no. 3, pp. 265-276, June 1995.

46. M.J.M. Pelgrom, and A.C. van Rens, "A 25 Ms/s 8-Bit CMOS ADC for embedded applications," *Proceedings 19th of European Solid State Circuits Conference*, pp. 13-16, 1993.

47. F.P.M. Beenker, "Testability concepts for digital ICs," *Ph.D. Thesis,* University of Twente, Netherlands, 1994.

48. M. Sachdev and Bert Atzema, "Industrial relevance of analog IFA: a fact or a fiction," *Proceedings of International Test Conference*, pp. 61-70, 1995.

49. R.J. van de Plassche, "Integrated analog-to-digital and digital-to-analog converters," *ISBN 079-2394-364,* 1994.

50. M. Mahoney, "DSP-Based testing of analog and mixed-signal circuits," *ISBN 0-8186-0785-8.*

51. R. Mehtani, B. Atzema, M. De Jonghe, R. Morren, G. Seuren, and T. Zwemstra, "Mix Test: a mixed-signal extension to a digital test system," *Proceedings of International Test Conference*, pp. 945-953, 1993.

52. M. Sachdev, "A DfT method for testing internal and external signals in A/D converters," European patent application, 1996.

53. B. Atzema, E. Bruls, M. Sachdev, and T. Zwemstra, "Computer-aided testability analysis for analog circuits," *Proceedings of the workshop on Advances in Analog Circuit Design*, April 1995.

54. E. Bruls, "Variable supply voltage testing for analogue CMOS and bipolar circuits," *Proc. Int. Test Conf.*, pp. 562-571, 1994.

55. R. Becker *et al.*, "PACT - A programmable analog CMOS transmission circuit for electronic telephone sets," *Proc. European Solid State Circuits Conf.*, pp 166-169, 1993.

56. M. Sachdev, *Defect Oriented Testing for CMOS Analog and Digital Circuits*, Kluwer Academic Publishers, Norwell, MA, 1997.

Fault Simulation

C.-J. Richard Shi

T his chapter provides an overview of fault simulation techniques for analog and mixed-signal (mixed analog and digital) circuits. Given an analog or a mixed-signal circuit, and a list of possible *faults*, the purpose of fault simulation is to find the response of a circuit when one fault or a cluster of faults occurs in the circuit. For example, consider an operational amplifier circuit. Possible faults, among others, may include a transistor gate open, the value of compensation capacitance shifted by half. The circuit responses that fault simulation computes can include DC operating points or simply DC responses (DC *fault simulation*), frequency-domain responses (*AC fault simulation*), and time-domain responses (*transient* or *time-domain fault simulation*). DC, AC, and transient fault simulation of analog and mixed-signal circuits are referred to as *analog fault simulation*.

3.1 Introduction

Analog fault simulation can be implemented in a straightforward manner by repeatedly injecting faults into a circuit and invoking a circuit simulator on the modified circuit. Early work along these lines includes FSPICE [48], ISPICE[25], and SFA[17]. These efforts are based on the public-domain SPICE simulator[36]. Recent examples of such an approach, are AnaFault[51] based on the ELDO simulator from Mentor Graphics, and a system based on the SABER simulator from Analogy[5,6]. This procedure is easy to implement (using scripts) and capable of exploiting the power of the underlying simulator and its environment. However, a naive implementation of this procedure using a circuit simulator is not very useful, for the following two reasons:

- *Numerical problem*: A fault may cause the circuit equations formulated by a circuit simulator to be ill-conditioned. This causes the circuit simulator to fail.
- *Computational cost*: Each analog circuit simulation, especially the time-domain simulation of nonlinear analog circuits, is known to be very time-consuming. Since each fault is treated separately, the number of circuit simulations required is at least equal to the number of faults. Further, unlike today's digital circuits, because of process variations and the uncertainty of operating environments and temperatures, parameters in analog and mixed-signal integrated circuits have a range of possible values. Since the circuit responses are highly sensitive to the changes in the parameter values, Monte Carlo simulation is commonly used to consider parameter variations[69]. As a result, the total number of circuit simulations can be prohibitively large.

The purpose of this chapter is to review:

1. Techniques developed specifically for analog fault simulation that do not use a circuit simulator in the inner loop
2. Techniques that use a circuit simulator intelligently by exploiting fault simulation specific features.

Only general-purpose fault simulation techniques, not those targeted at specific types of circuits, are explored. We assume that the reader is familiar with basic concepts and techniques in analog circuit simulation[1]. We note that the techniques to speed up standard circuit simulation such as tearing[84], relaxation methods[40], Elmore-delay-based simulation methods[54,55,56], AWE[44], and Block-Lanczos algorithms[20], can be used directly to speed up analog fault simulation. These techniques, including those developed in the context of fault simulation[78], will not be reviewed here.

3.1.1   Why Analog Fault Simulation?

As discussed in Chapter 1, there has been a significant spurt in activity in analog and mixed-signal circuit test. A fault-driven and simulation-based methodology for analog and mixed-signal test, based on the paradigm used to test digital circuits,

---

1.   Nagel's Ph.D. dissertation written about a quarter century ago still reflects the state of the art in most circuit simulators! Two excellent textbooks on circuit simulation were written by McCalla[31], and Vlach and Singhal[79]. The book by Kundert[26] covers many practical aspects of circuit simulation.

has attracted a lot of attention[3,11,12,19,35,34,37,42,47,68,69,76]. In this method-ology, analog circuits are tested on whether they are free of (manufacturing and/or design) faults. By using automatic fault simulation and test generation tools, such a methodology can use less expensive tests, reduce test time, and have a shorter test development cycle. When combined with built-in-self testing (BIST), this methodol-ogy promises to address the challenges of mixed-signal testing. Analog fault simula-tion will be a cornerstone tool for the industry wide use of a fault-driven analog test methodology. Similar to the use of fault simulation in digital testing, analog fault simulation can be used for fault coverage analysis, fault grading, fault collapsing, and BIST. Analog fault simulation may play an even more significant role in fault-driven analog testing since, from our experience, simulation-based test generation appears to be the most viable approach to test generation for fault detection and diagnosis of mixed-signal systems.

Analog fault simulation also provides an important computer-aided design tool for analog designers. Unlike digital design, no viable synthesis tools exist for analog circuits. Analog design today is primarily done manually by experienced analog designers. One of the widely used analyses that designers rely on most in the design phase is *what-if analysis*. For example, an analog designer may ask the fol-lowing what-if questions: What will happen if the value of a particular resistor dou-bles? What will happen if a particular current mirror is switched to another current mirror? What will happen if two wires are routed close to each other? Unfortunately, what-if analysis is done today in almost the same way as was done 20 years ago, manually, or using a standard circuit simulator. We observe that these what-if ques-tions can be represented as variations to the circuit and can be solved by analog fault simulation. Hence efficient and effective fault simulation methods are useful for analog design itself. Analog fault simulation can also help solve the related problem of analog design diagnosis. Frequently, when a design fails to meet the specification, we may ask what part of a design can be changed to meet the given performance specification.

Lastly, analog fault simulation can potentially be used to test digital circuits. With the constant progress in microelectronic manufacturing technologies, the dimension of transistors in digital integrated circuits is continually shrinking, and the clock cycle time used in digital integrated systems is quickly approaching the GHz range. This has two implications for digital testing, that are relevant to us. First, traditional digital testing based on 0-1s is no longer sufficient. Bridging faults, delay faults, and noise all become important. It will not be surprising if these analog phe-nomena dominate the behavior of digital systems in year 2010 when the feature size is predicted to be 0.05 micron[39]! Secondly, due to the enormous number of 0-1 testing vectors required, the inadequacy of digital testers in handling high-frequency digital testing, and the limited number of pins available for testing may make a traditional approach untenable. High-level digital testing, that is test based on the functionality of the system as a whole, is seen as an attractive alternative to traditional 0-1 testing. Many high-level abstractions of digital systems are analog. In fact, some test engineers have already successfully used analog and mixed-signal testers for some digital testing.

3.1.2   Analog Fault Models and What-if Analysis

Analog fault modeling is a fundamental, yet still controversial issue, in fault-driven analog testing. We believe that part of the controversy comes from the attempts to borrow and to interpret the concept of a fault in the digital world, and apply it to the analog world. In this subsection, we attempt to provide a slightly different perspective. An analog circuit under test can have three testing outcomes:

- *Catastrophic (hard) failure*: The circuit is not functioning at all. For example, an oscillator does not oscillate. A typical result is that the output node is stuck at a particular voltage level (for example, 0 Volts, 5 Volts, or an intermediate voltage).
- *Unacceptable performance degradation*: In this case, the circuit is still functioning, but some of its performance specifications lie outside their acceptable ranges. For example, an oscillator may still oscillate but at a very low frequency. Usually, performance acceptance (or tolerance) ranges are determined by many factors such as the application environment and various economic considerations. Performance degradation is usually referred to as a soft failure.
- *Acceptable performance*: The circuit is functioning and all its performance parameters are within their specification ranges. In this case, the circuit is said to be *correct*.

We note that catastrophic failures can be detected by DC testing or power supply monitoring[67], and parameter degradation is more difficult and costly to detect.

We can classify all those *changes to the circuit* that cause the circuit to fail catastrophically as *catastrophic (hard) faults*, and those changes that cause performance degradation as *parametric (soft) faults* or *parametric degradations*. This definition is different from the definition traditionally used to define faults in analog and mixed-signal circuits. Traditionally, catastrophic faults and parametric faults are defined with respect to the local failure mechanism[4,15,30], not their impact on the circuit. Often, these two definitions are equivalent. But we feel that the traditional definition is conceptually confusing, due to the following analog-specific reasons:

- For analog integrated circuits, parameters are always associated with certain variations, from their nominal values. Only those variations that cause the performance of the circuit to lie outside the specification ranges are real faults. Therefore, the traditional definition of parametric faults as the variation from its nominal value (regardless of its actual impact on the circuit performance) is misleading to analog designers.

- As has been observed by several researchers[7,58,64] local catastrophic faults in analog circuits such as opens and shorts do not necessarily lead to the catastrophic failure of the entire circuit. Sometimes, they cause only parametric or soft failures. Thus, such local catastrophic faults can be represented as parametric faults at a higher level of abstraction.

From the perspective of physical failure mechanisms, changes to a circuit come from a wide range of sources such as manufacturing defects[13], process variations[13], circuit and environmental parasitics, changes in the environment or ambient temperatures, and design errors/nonrobustness. Only those physical failures that cause the circuit performance parameters to lie outside the specification ranges and/or those that may affect circuit reliability need to be detected and classified as real faults. Usually, as discussed in Chapter 2, physical defects can be derived from process data and circuit layouts using inductive fault analysis[81,83]. Parasitics can be estimated or extracted using extraction tools. For the purpose of analog testing, any possible change to a circuit can be analyzed and examined based on its impact on the circuit's performances and reliability. An efficient and effective tool to do this is analog fault simulation. However, faults here are various what-if situations. From the simulation perspective, faults are changes to a circuit. How they are extracted, whether they are realistic, and how they impact the circuit performances and reliability are irrelevant. Therefore, perhaps a better terminology for analog fault simulation is "analog what-if simulation."

Similar to digital circuits, analog and mixed-signal circuits can be described at various *abstraction levels*: For example, at the transistor-level (nonlinear differential equations) and at the functional-level (signal-flow graphs, transfer functions). At each level, there are two views: *structural* and *behavioral*. For example, a transistor-level netlist is the structural view of a circuit, and nonlinear differential equations are the corresponding behavioral view. Associated with each view, there can be many models[57,61]. Consequently, faults can be the changes to the descriptions at various abstraction levels in either of two views. From this perspective, we can classify faults into the following four categories:

- (*Simple*) *structure faults*: Changes to the description in the structural view. At the transistor level, examples of structure faults include device opens and shorts, and interconnect opens and shorts. Equivalently, we can define structure faults at the signal-flow level.

- *Parameter faults*: Changes of parameter values that cause the circuit performances to lie outside their performance specifications. At the transistor level, examples of such parameters include resistances, capacitances, transistor channel length and width, etc.
- *Matching faults*: Device and structure matching is an important type of analog design constraint. Usually two or more components or two parts of the circuit are required to match exactly with respect to their values or structures in order to achieve the performance requirements (less sensitive to process variations). We refer to the failure of meeting matching requirements as matching faults.
- *Delta faults*: This class includes any changes of the circuit. These could range from a local change such as the addition of a parasitic network associated with a wire, the misuse of certain components, and to global changes such as a dramatic change to a distinctly different circuit. Conceptually, we can view structure faults and parameter faults as special cases of delta faults.

We note that traditionally, analog fault simulation has only targeted transistor-level circuits. This includes modeling shorts, opens, stuck-at faults, soft faults, inversion, and replace faults as described in[6].

There are two means of describing an analog circuit: graphically via schematic capture or textually via hardware description languages. Correspondingly, faults can be concretely represented and injected based on schematics or hardware-description languages. An example of faults injected into a hardware-description language description can be found in[2]. Since a schematic usually represents the actual physical implementation, schematic-induced faults are generally more realistic than language-induced faults[41].

Process variations can be described either statistically or as worst-cases. In general, process variations are of a statistical nature. As a consequence, we feel that faults, yields, testability, and fault coverage for analog integrated circuits are more appropriately defined statistically using the notion of probability. Nevertheless, in this chapter, we choose to concentrate on fault simulation methods based on worst-case modeling for the following considerations: (1) process statistics are usually private information confined in the foundry and are not available to analog circuit designers or to the public; (2) analog designers are not statisticians; (3) statistical modeling is an emerging and much less mature research direction in mixed-signal testing.

3.1.3   Focus and Organization

This chapter reviews efficient and effective fault simulation techniques for analog and mixed-signal circuits from the algorithmic perspective. Early work was surveyed in [4,15] and in a book edited by Liu[30]. Excluded from this survey are discussions about: (1) how faults are extracted; (2) how faults are modeled and described; (3) techniques developed for digital circuits or specific types of analog circuits. Recent work in analog fault modeling and automated analog test generation has been reviewed elsewhere[65,66]. Attempts have been made to be subjective; however, this chapter may be mildly biased and narrowed towards research work done in our group. This chapter is organized as follows. We first review the techniques developed for linear analog circuits in Section 3.2. Section 3.3 describes techniques for the DC fault simulation of nonlinear analog circuits. Section 3.4 covers general techniques which are useful for time-domain simulation of large mixed-signal circuits. Section 3.5 concludes this chapter.

3.2 Fault Simulation of Linear Analog Circuits

Linear analog circuits constitute a large class of mixed-signal circuits and systems that are widely used in video and image processing, digital signal processing, control, communications, and many other applications. Although implemented using transistors, which are essentially nonlinear devices, and passive components, such circuits operate in the linear region under normal conditions. They can be modeled by small-signal models linearized at the *operating point* as far as AC characteristics are concerned. "Catastrophic faults" in such circuits may change the operating points and invalidate circuit linearity. However, such faults are generally easy to detect. The most "difficult-to-detect" faults are those that cause slight behavior changes. Therefore, efficient fault simulation of linear analog circuits is itself a practical interesting problem. On the other hand, the "linearity" of the circuit enables the development of efficient algorithms. Research in this area is mature and even finds application to nonlinear fault simulation and Monte Carlo analysis.

3.2.1 Householder's Formula

A linear time-invariant circuit can be represented, using the modified nodal analysis method[23], by the following set of system equations:

$$Tx = w \tag{3.1}$$

where T is the circuit matrix, x is the vector of circuit unknowns (node voltages and certain branch currents), and w is the contribution of voltage and current sources[79]. The purpose of fault simulation is to simulate the circuit with a given list of faulty conditions. Each faulty condition may be composed of more than one faulty element (the resulting circuit is called *faulty circuit*). We can write the set of system equations for a faulty circuit as

$$T_f x_f = w_f \tag{3.2}$$

where T_f and w_f represent the matrix and the right-hand-side vector respectively, for the faulty circuit.

A fundamental observation for efficient fault simulation is that only a single component or a few components may fail simultaneously. Thus, the circuit matrix T_f for the faulty circuit differs in only a small number of entries from the good-circuit matrix. As a result, Householder's formula can be applied to obtain the exact solution of the faulty circuit efficiently without resolving the whole set of equations.

Let A be an $n \times n$ matrix, U be an $n \times m$ matrix, S be an $m \times m$ matrix, and W be an $m \times n$ real matrix, then Householder's formula states[24]:

$$(A + USW)^{-1} = A^{-1} - A^{-1}U(S^{-1} + WA^{-1}U)^{-1}WA^{-1}$$

Householder's formula was used for circuit simulation in the 1970s in the area known as large-change sensitivity. It was first applied by Temes to analog fault simulation[72]. Vlach and Singhal proposed a method for efficient fault simulation of linear circuits without using Householder's formula explicitly. However, it yields the same numerical procedure as does the direct use of Householder's formula [79].

We consider how Householder's formula can be applied to the simulation of a faulty circuit, in which m components are faulty. For the most part, we can restrict the components subject to changes to be the types that enter the modified nodal equations in the form of $t_i p_i q_i^T$ with

$$p_i = e_{i_j} - e_{i_{j'}}$$
$$q_i = e_{i_k} - e_{i_{k'}}$$

and with e_l being a vector of zeros except for the l-th entry, which is equal to 1.

As mentioned above, assume that there are m components that are subject to large changes by $h_1, h_2, \ldots, h_m$ and let them change by the amounts $\delta_1, \delta_2, \ldots, \delta_m$. Then

$$T_f = T + \sum_{i=1}^{m} \delta_i p_i q_i^T$$

or

$$T_f = T + P\Delta Q^T$$

Here P and Q are $n \times m$ matrices which contain 0 and ± 1 entries:

$$P = [p_1, p_2, \ldots, p_m]$$

and

$$Q = [q_1, q_2, \ldots, q_m]$$

and Δ is a diagonal $m \times m$ matrix:

$$\Delta = diag[\delta_i]$$

To apply Householder's formula, let $A = T$, $U = P$, $S = \Delta$, and $W = Q^T$, then we have

$$T_f^{-1} = T^{-1} - T^{-1}P(\Delta^{-1} + Q^T T^{-1} P)^{-1} Q^T T^{-1}$$

Hence,

$$x_f = (T^{-1} - T^{-1}P(\Delta^{-1} + Q^T T^{-1}P)^{-1}Q^T T^{-1})w_f$$
$$= T^{-1}w_f - T^{-1}P(\Delta^{-1} + Q^T T^{-1}P)^{-1}Q^T T^{-1}w_f$$

We consider the use of this formula in a SPICE-like simulator[2]. The matrix T^{-1} is usually not directly available. But as the result of the simulation of the good circuit, the LU factors of T are known. Then $T^{-1}w_f$ can be obtained by one forward substitution and one back substitution. The cost is n^2. Let its result be y. Then we note that $Q^T y$ is a sparse vector of size m. It can be obtained by inspection, and no long operation is needed. Matrix $T^{-1}P$ can be obtained by using the LU factors of T and by performing m forward and back substitutions, with the columns of P serving, in turn, as right-hand-side vectors. The cost of this operation is mn^2. Then $Q^T(T^{-1}P)$ is a matrix entry selection operation. It can be obtained by inspection, and no long operation is needed. The computation of Δ requires m operations. Now consider the computation of $(\Delta^{-1} + Q^T T^{-1}P)^{-1}(Q^T y)$. Let its result be z. This is equivalent to solving the following equation:

$$(\Delta^{-1} + Q^T T^{-1}P)z = Q^T y$$

Since the matrix $(\Delta^{-1} + Q^T T^{-1}P)$ and the vector $Q^T y$ are known, the number of operations required to solve this equation is:

$$\frac{m^3}{3} + m^2 + \frac{m}{3}$$

Since computing $T^{-1}Pz$ takes nm operations, the total number of operations for simulating the faulty circuit containing m faulty components is:

$$n^2 + mn^2 + \frac{m^3}{3} + m^2 + mn + \frac{2}{3}m$$

We note that for many fault simulation runs, it is advantageous to have T^{-1} computed explicitly. This matrix can be obtained by using the LU factors of T and performing n forward and back substitutions, with the columns of I serving, in turn, as right-hand-side vectors. The cost is n^3. Then both $T^{-1}P$ and $Q^T(T^{-1}P)$ can be obtained directly by inspection. Suppose that there are η fault simulation runs in total, the number of operations for simulating each faulty circuit is:

$$\frac{n^3}{\eta} + n^2 + \frac{m^3}{3} + m^2 + mn + \frac{2}{3}m$$

In practice, $m \ll n$. As described by Temes, for a circuit with 30 to 50 components, fault simulation by using Householder's formula is at least 1000 times

---

2.   For more information about using LU decomposition to solve a set of linear equations, the reader is referred to[31,79] for details.

faster than repeatedly solving the equations and 10 times faster than sparse-matrix-technique-based repeated analysis.

In most situations, only a single faulty component is considered in one fault simulation, and $w_f = w$. Then the procedure based on Householder's formula can be simplified significantly. In this case, we have

$$T_f^{-1} = T^{-1} - \frac{T^{-1} e_i e_j^T}{(\Delta t_{ij})^{-1} + (T^{-1})_{ji}} T^{-1} \tag{3.3}$$

Note that

$$x_f = T_f^{-1} w \text{ and } x = T^{-1} w \tag{3.4}$$

then we have

$$x_{fk} = x_k - \alpha_k x_j \text{ and } k = 1,2,\dots,n \tag{3.5}$$

where

$$\alpha_k = \frac{[(T)^{-1}]_{ki}}{(\Delta t_{ij})^{-1} + (T^{-1})_{ij}} \tag{3.6}$$

The equation above can be further simplified for each specific type of fault:

• Short fault: Suppose that a fault forces the value of matrix entry T_{ij} to be zero. Then $\Delta t_{ij} = -T_{ij}$, and

$$\alpha_k = \frac{[(T)^{-1}]_{ki}}{-T_{ij}^{-1} + (T^{-1})_{ij}} \tag{3.7}$$

• Open fault: Suppose that a fault forces the value of matrix entry T_{ij} to be infinity, then $\Delta t_{ij}^{-1} = 0$ and we have

$$\alpha_k = \frac{(T^{-1})_{ki}}{(T^{-1})_{ij}} \tag{3.8}$$

In addition to the speed advantage, the use of Householder's formula avoids the numerical problems associated with direct simulation of faulty circuits[29,42,80]. Although this approach is not directly applicable to time-domain or nonlinear DC fault simulation, it has been exploited for fault diagnosis[9], for efficient band-fault simulation[42], for interval-mathematics-based band computation[59], frequency-domain fault simulation[73], and nonlinear DC fault simulation[60].

3.2.2 Discrete Z-domain Mapping

Nagi, Chatterjee, and Abraham [38] proposed a simulation technique based on mapping the good and faulty circuits to the discrete Z-domain. Their method starts by representing the circuit using state equations in the complex frequency s-domain:

$$sX(s) = AX(s) + BU(s)$$

These state equations can be derived by using signal-flow graph techniques and using dummy state variables.

Then integration is approximated by using recursive differences to obtain a bilinear transformation of s-domain equations to the Z-domain. These equations can be solved to obtain a discretized (sampled) solution.

The bilinear transform is given by

$$s = \frac{2}{t_s}\frac{z-1}{z+1}$$

where t_s is the sampling time. With this, the equation set above can be transformed to

$$X(z) = Z_A z^{-1} X(z) + Z_B [z^{-1} u(z) + u(z)]$$

Here z^{-1} represents a single delay. This equation set can be rewritten in the time domain as:

$$x(t_k) = Z_a X(t_{k-1}) + Z_B (u_{k-1} + u_k)$$

A given input signal $u(t)$ is sampled at the rate $1/t_s$.

With this method, the authors have demonstrated an order of magnitude speedup over a traditional circuit simulator[38]. The speedup comes mainly from the use of simple behavioral models for operational amplifiers. However, the process of mapping a fault to the Z-domain is nontrivial: (1) A single fault gives rise to multiple faults; (2) The number of states could be changed by the fault; (3) The use of the state variable approach is not desirable; (4) The transformation is generally restricted to certain circuit classes; (5) The choice of a sampling frequency is an issue; Nevertheless, fault mapping to the Z-domain remains an interesting idea.

3.2.3 Fault Bands and Band Faults

Pahwa and Rohrer were the first to consider fault simulation of linear analog circuits under component parameter variations[42]. Their approach to computing the worst-case response bands consists of three steps:

- The sensitivity with respect to each (component) parameter is computed by using one forward substitution and backward substitution from the **LU** factors of nominal circuit simulation.
- The corner cases are identified by assuming that the worst-case circuit response band of the good circuit is computed by assuming all the components

are at their extreme parameter boundaries in such a direction so as to reinforce the deviation of the nominal response.

• The worst-case response bounds are computed by solving the circuit equations at the corners. This requires two LU decompositions and two forward and backward substitutions.

In total, the computation of the worst-case bound for the good circuits takes two LU decompositions and 3 forward and backward substitutions.

Then, the faulty response bands can be computed in two different ways as illustrated in Figure 3-1. In the first approach, the nominal response of the faulty circuit is first computed using Householder's formula. This requires 1 forward and backward substitution. Next, the sensitivity is computed at the nominal point using 1 forward and backward substitution. Then, two LU decompositions and 2 forward and backward substitutions are required to solve the two worst-case equations. Therefore, for each fault, a total of 2 LU decompositions and 4 forward and backward substitutions are required.

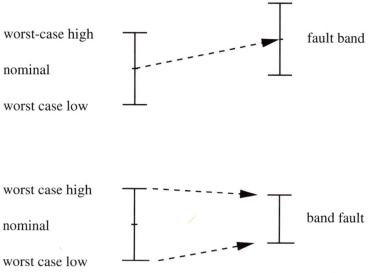

Figure 3-1   *Illustration of band faults.*

In the second approach, it is assumed that the worst cases of the faulty circuit occur at exactly the same corner as the worst cases of the good circuit. Then for a faulty circuit, the worst-case high and worst-case low can be computed via Householder's formula from that of the good circuit. This involves one forward and backward substitution for each bound of each fault. So the total computational effort is only *2m* forward and backward substitution for *m* faults.

These approaches are computationally efficient. However, there exist two deficiencies. First, it is well known that the worst-case response bands estimated as discussed above may lead to overly pessimistic results[79]. On the other hand, there is no guarantee that the computed band encloses the actual band. Second, the assumption that the worst-case response occurs at the same extreme parameter boundaries for both the faulty-free and faulty circuits is not true in general.

3.2.4   Interval-Mathematics Approach

Recently, an approach based on interval mathematics has been developed to handle fault simulation under parameter variations that removes the two limitations in the approaches of Pahwa and Rohrer[42]. The method uses slightly more CPU time than the band-fault approach, but its result is guaranteed to be correct, the bound always contains the actual bound, and further the difference is usually small.

3.2.4.1   Basic Notations

Let $p \in \Re$ be a real number whose *value* may not be precisely known. Instead, we are often given a range (value set) and the exact value of p is uncertain within this range. This can be represented by an *interval number* p^I, with a *lower (left) bound* p^L and an *upper (right) bound* p^R, denoted by $p^I = [p^L, p^R]$. The *midpoint* $mid(p^I)$ of an interval number p^I is defined as:

$$mid(p^I) = \frac{1}{2}(p^L + p^R)$$

and the *radius* $rad(p^I)$ of p^I is defined as:

$$mid(p^I) = \frac{1}{2}(p^R - p^L)$$

An *interval vector* x^I is a vector whose elements are interval numbers, and we write an interval vector as $x^I = [x^L, x^R]$. An *interval matrix* A^I is a matrix whose elements are interval numbers and we write an interval matrix as $A^I = [A^L, A^R]$.

A *system of linear interval equations:*

$$A^I x = b^I$$

with coefficient matrix A^I and right-hand-side vector b^I is defined as the collection of systems of linear equations:

$$\{Ax = b | A \in A^I \text{ and } b \in b^I\}$$

Its *solution set* is given by:

$$S = \{x = A^{-1}b | A \in A^I \text{ and } b \in b^I\}$$

The *solution hull* of a system of linear interval equations is defined as the tightest interval hull (vector) enclosing S. The projection of a solution hull into a particular dimension is an interval. With a slight abuse of notation, we sometimes use x^I to denote the solution hull, and x^R and x^L are called *solution bounds*. The system of linear interval equations can be written as:

$$A^I x^I = b^I \tag{3.9}$$

Given two interval numbers a^I and b^I, the following interval arithmetic operations are defined:

$$a^I + b^I = [a^L + b^L, a^R + b^R]$$
$$a^I - b^I = [a^L - b^R, a^R - b^L]$$

$$a^I \bullet b^I = [min(a^L b^L, a^L b^R, a^R b^L, a^R b^R), max(a^L b^L, a^L b^R, a^R b^L, a^R b^R)]$$

$$a^I / b^I = [min((a^L/b^L), (a^L/b^R), (a^R/b^L), (a^R/b^R)),$$
$$max((a^L/b^L), (a^L/b^R), (a^R/b^L), (a^R/b^R))]$$

For example, let $X^I = [0,2]$, then $1 - X^I = [-1,1]$, $X^I \bullet X^I = [0,4]$, and $1 - X^I + X^I \bullet X^I = [-1,1] + [0,4] = [-1,5]$, where the actual value of the set $\{1 - x + x \bullet x | x \in X^I\} = [3/4,3]$. This demonstrates a peculiar characteristic of interval operation: the value set may be *expanded* (*overestimated*) due to the fact that *correlations* among the values represented by intervals (e.g., $1 - X^I = [-1,1]$ and $X^I \bullet X^I = [0,4]$) are ignored by interval arithmetic. In general, key challenges in effectively applying interval analysis to real applications are: (1) to reduce the amount of correlations among interval terms, and (2) to reduce interval operations as much as is possible.

3.2.4.2   Equation Formulation of the Circuit under Test

Since the quality of interval analysis depends (often crucially) on how the equation is formulated. We propose to use a variant of Modified Nodal Analysis (MNA) to formulate the equations for the circuit under test. The formulation is particularly amenable to interval analysis. To motivate the formulation, we consider the following example.

Consider the circuit shown in Figure 3-2, where r_1 is a parameter with variations. The following formulation of circuit equations is used in most circuit simulators such as SPICE[36]:

$$\begin{bmatrix} \dfrac{1}{r_1^I} & -\dfrac{1}{r_1^I} \\[2ex] -\dfrac{1}{r_1^I} & \dfrac{1}{r_1^I} + \dfrac{1}{r_2^I} \end{bmatrix} \begin{bmatrix} v(1) \\ v(2) \end{bmatrix} = \begin{bmatrix} 1 \\ 0 \end{bmatrix}$$

Since r_1^I appears in all the four entities of the coefficient matrix, these four entries are correlated. To resolve this problem, we can introduce another variable:

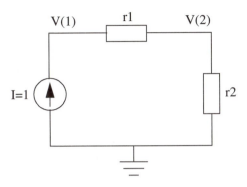

Figure 3-2 *A simple two resistor circuit.*

the branch current through resistor r_i^I. Then the circuit equation set can be reformulated as:

$$
\begin{bmatrix} 0 & 0 & 1 \\ 0 & \dfrac{1}{r_2} & -1 \\ 1 & -1 & -r_1^I \end{bmatrix}
\begin{bmatrix} v(1) \\ v(2) \\ i(r_1) \end{bmatrix} =
\begin{bmatrix} 1 \\ 0 \\ 0 \end{bmatrix}
$$

Here all the interval elements in the coefficient matrix are independent. The idea above can be generalized by changing slightly the Modified Nodal Analysis (MNA) method developed by Ho et. al.[23] and popularized by SPICE[36]. Basically, for all the components which do not have interval parameters, the rules of MNA are followed. For every component with an interval parameter, an additional variable (branch current) and an additional equation (branch equation) are introduced. In general, for a linear time-invariant circuit with some circuit parameters represented by interval numbers, applying these rules leads to a system of complex linear interval equations in the frequency domain represented as:

$$
(\mathcal{G}^I + s\mathcal{C}^I)x = w^I \tag{3.10}
$$

where x is the unknown vector composed of nodal voltages and branch currents resulting from both MNA rules and the rule above for interval parameters, the matrix $\mathcal{G}^I$ stores all conductors and frequency-independent numbers arising in the formulation, the matrix $\mathcal{C}^I$ stores all capacitors and inductors values and other values that are associated with the frequency variable s, and w^I represents contributions from current sources, voltage sources, as well as initial conditions associated with capacitors and inductors. For convenience, we refer to this variant of the MNA formulation as a *generalized MNA formulation*, or simply *Generalized Nodal Analysis (GNA)*. Note that the GNA formulation for a circuit without interval parameters degenerates to the MNA formulation.

Frequency-domain simulation is performed by substituting $s = j\omega$ in Eqn. 3.10 and solving it for a given frequency range. Let $x = x_{\mathcal{R}} + jx_{\mathcal{I}}$ and $w^I = w^I_{\mathcal{R}} + jw^I_{\mathcal{I}}$, where subscripts $\mathcal{R}$ and $\mathcal{I}$ denote, respectively, the real part and the imaginary part of a complex vector. Then Eqn. 3.10 can be rewritten as a system of *real* interval equations as follows:

$$\begin{bmatrix} \mathcal{G}^I & -\omega\mathcal{C}^I \\ \omega\mathcal{C}^I & \mathcal{G}^I \end{bmatrix} \begin{bmatrix} x_{\mathcal{R}} \\ x_{\mathcal{I}} \end{bmatrix} = \begin{bmatrix} w^I_{\mathcal{R}} \\ w^I_{\mathcal{I}} \end{bmatrix} \tag{3.11}$$

3.2.4.3   Solving Systems of Interval Linear Equations

Let C be the *midpoint* of interval matrix A^I, i.e., $C = \frac{1}{2}(A^L + A^R)$. Premultiplying Eqn. 3.9 with C^{-1} transforms the *original system* into a new system:

$$M^I x = r^I \tag{3.12}$$

where $M^I = C^{-1}A^I$ and $r^I = C^{-1}b^I$. This transformation process is called *preconditioning*, and Eqn. 3.12 is called the *preconditioned system*. With a set of theoretical characterizations, an elegant algorithm has been derived in[59]. The algorithm is summarized in Figure 3-3.

3.2.4.4   Concurrent Fault Simulation

Fault simulation can be performed by reconfiguring the circuit, and applying the above algorithm directly on the faulty circuit. Since the M^L matrix for the faulty circuit differs only slightly from that of the good circuit, the matrix P can be computed efficiently using Householder's formula. Therefore, approximately two forward and backward substitutions are required for each fault simulation.

3.2.5   Summary

In this section, we have reviewed several methods for the efficient fault simulation of linear analog circuits. For fault simulation without process variations, the use of Householder's formula is the first developed yet the most effective method. To consider process variations, both the band-fault approaches and the interval-mathematics approach can be used. The interval-mathematics approach is capable of producing very tight and guaranteed correct response bounds but with slightly more CPU time than the band-fault approaches. For time-domain fault simulation, either the approach based on frequency-domain Householder's formula combined with FET or the approach of Z-domain fault mapping can be used.

Interval System_Solve (A^I, b^I)

1. $C \leftarrow \frac{1}{2}(A^L + A^R)$
2. $M^I \leftarrow C^{-1}A^I$; $r^I \leftarrow C^{-1}b^I$
3. $P \leftarrow (M^L)^{-1}$
4. for $i = 1$ to n do
5. $s_i \leftarrow \left| mid(r_i^I) \right| + rad(r_i^I)$
6. for $i = 1$ to n do

7. $f_i \leftarrow \sum\limits_{j=1}^{n} P_{ij}s_j$

8. $g_i \leftarrow f_i - 2P_{ii}\left| mid(r_i^I) \right|$
9.   if $g_i \geq 0$
10.   if $mid(r_i^I) > 0$
11. $x_i^I \leftarrow [-g_i, f_i]$
12.   else
13. $x_i^I \leftarrow [-f_i, g_i]$
14.   else
15.   if $mid(r_i^I) > 0$
16. $x_i^I \leftarrow [-g_i/(2P_{ii} - 1), f_i]$
17.   else
18. $x_i^I \leftarrow [f, -g_i/(2P_{ii} - 1)]$
19. return x^I

Figure 3-3 *An algorithm for solving systems of linear interval systems $A^I x = b^I$.*

3.3 DC Fault Simulation of Nonlinear Analog Circuits

When compared to several analog test methods, DC testing requires less expensive test equipment and shorter testing times, and it is therefore preferred. Efforts have been devoted to exploit low-cost DC test generation[34] and DC built-in self-test[8]. In the testing scenario where DC testing is performed before complex AC functional testing and transient testing, DC tests that detect the majority of faults would reduce the overall testing time and cost.

DC fault simulation is a tool to analyze how well a DC test can detect a given list of faults. That is, it is a tool for DC-fault coverage analysis. It is also important for DC-test generation. DC fault simulation is a time-consuming and challenging task. The root of the difficulty is that DC simulation is usually done by solving a set of nonlinear equations as in SPICE-like simulators, which may require many iterations to converge. The strong nonlinearity arising in modeling open/short faults often causes a numerical problem to a simulator. Consequently, the simulator either fails to converge, or simply fails due to the matrix-singularity problem. In this section, we review several techniques developed to address both the CPU time and convergence problem of DC fault simulation of nonlinear analog circuits.

3.3.1   The Complementary Pivot Method

An interesting approach taken by Lin and his students[29,70] to DC fault simulation is based on the complementary pivot theory[18,28]. It is observed that the demand of accuracy in DC fault simulation is less severe than the usual accuracy requirements in circuit simulation. This observation lays down the foundation to model the circuit nonlinearity by ideal diodes. This, combined with fault modeling using switches, leads to a formulation that is particularly amenable to a solution method in operational research called complementarity pivoting.

Being distinct from the widely-used Newton-Raphson algorithm for DC analysis, the complementary pivot method does not suffer the same numerical problem caused by the extreme (nonsmooth) nonlinearity arising in the modeling of analog circuits and associated faults. It consists of the following four major steps:

1. Modeling all the nonlinear devices by piecewise-linear I-V characteristics, in particular by ideal diodes;
2. Using switches to represent open, short, and parametric faults;
3. Formulating the fault simulation problem as the complementarity problem using n-port network theory;
4. Solving the resulting complementarity problem using Lemke's complementarity pivoting algorithm.

3.3.1.1   Modeling Nonlinear Analog Circuits with Ideal Diodes

An ideal diode is a 2-terminal device that behaves as an open circuit when reverse-biased and as a short circuit when forward-biased. Theoretically, models for nonlinear analog circuits can be constructed in such a way that all the nonlinearities are associated with 2-terminal nonlinear resistors[10]. A 2-terminal nonlinear resistor can be represented by a piecewise linear (usually abbreviated PWL) I-V curve to any desired accuracy by increasing the number of straight-line segments. In practice, the efficiency of the method depends largely on the number of segments used. Fortunately, for fault simulation where the demand of accuracy is less severe than in circuit simulation, a 3-segment or even 2-segment approximation to a diode I-V curve is adequate.

3.3.1.2   Modeling Faults with Ideal Switches

An ideal switch is an open circuit when it is turned off and a short circuit when it is turned on. Open, short, and parametric faults can all be modeled by ideal switches, as illustrated in Figure 3-4. The switches in Figure 3-4(a), Figure 3-4(b), Figure 3-4(c), are normally closed (NC), normally open (NO), and normally open

(NO) respectively. All the configurations represent a resistor with value R under normal conditions. Under faulty conditions, they represent, respectively, an open fault, a short fault, and a parametric fault (the value of R is halved).

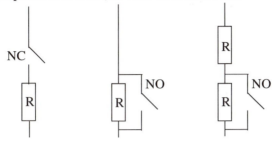

(a) Open fault   (b) Short fault   (c) Parametric fault

Figure 3-4   *Circuit faults injected using switches.*

3.3.1.3   *Formulating the Multiport Constraint Equation*

The circuit under test is modeled with linear resistors, controlled sources, d.c. independent sources, switches, and ideal diodes. Suppose that there are d diodes and m switches. Let there be k test nodes whose voltages are to be measured, and l branches whose currents are to be monitored. We view theses voltages v_k and currents i_l as those associated with k zero-valued independent current sources and l zero-valued independent voltage sources. A key observation is that these $k+l$ *measurement ports*, together with d diodes and m switches (each is itself a port), lead to $k+l+d+m$ pairs of so-called *complementarity* variables (port current and voltage), i.e., for each port, $v \geq 0$, $i \geq 0$ and $iv = 0$! This is illustrated in Figure 3-5.

With these $k+l+d+m$ ports extracted out, we can formulate the set of hybrid circuit equations for the remaining linear circuit in the following form known as the *complementarity problem*:

$$w = Mz + q \qquad\qquad (3.13)$$
$$w \geq 0 \qquad z \geq 0$$
$$w^T z = 0$$

where M is a square matrix of order $n = k+l+d+m$, and w, z and q are all vectors of $n=k+l+d+m$ components. The vectors w and z are composed of $2n$ port currents and voltages. They are the circuit unknowns to be found. The matrix M and vector q are known from the hybrid formulation. We note that the original work of Lin and Elcherif relies on the existence of the hybrid representation. However, this limitation has been removed recently by Vandenberghe, de Moor, and Vandewalle[77].

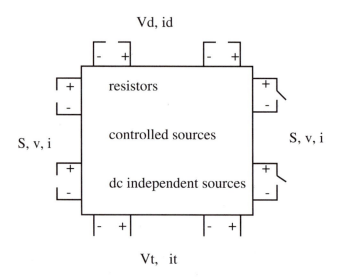

Figure 3-5 *Formulation of a resistive n-port.*

3.3.1.4   Solving the Complementarity Problem

In the original work of Lin and Elcherif[29], the resulting complementarity problem was solved using Lemke's pivoting algorithm[28]. Algorithms capable of finding all DC solutions more efficiently can be found in [77]. However, Lemke's algorithm still remains one of the most popular algorithms for solving the complementarity problem.

Lemke's pivoting algorithm is a homotopy method that creates a path in the solution space, leading from an initial point to a solution of the problem. It is based on adding an extra variable z_0 to the complementarity problem to form the following *expanded* problem:

$$w = Mz + z_0 e + q$$
$$w \geq 0 \qquad z \geq 0 \qquad s_0 \geq 0$$
$$w^T z = 0$$

where $e \in R^n$ and $e = (1, 1, \ldots, 1)^T$. A solution to the expanded problem can be obtained by inspection: let $z=0$ and z_0 be equal to the absolute value of the most negative component of q. Then $w = z_0 e + q \geq 0$.

A set of n variables, out of the $2n$ variables of w and z, that are not equal to zero forms the *basic set* and the remaining variables are called *nonbasic variables*. An intuitive observation is that if there exists a solution, then we can always reorganize Eqn. 3.13 and the associated $2n$ variables in such a way that the basic set of variables appear in the left-hand side of Eqn. 3.13, and the non-basic variables appear in the right-hand side. That is, such that the corresponding columns of

$[I - M]$ form a nonsingular matrix. Then a solution can be obtained by setting all the non-basic variables to zero and solve Eqn. 3.13 for the values of the basic variables. This observation is the basis of the tableau method described below.

Note that a solution to the modified problem with $z_0 \neq 0$ is not a solution to the original problem! Since z_0 is a basic variable, then the corresponding non-basic set must contain a complementary pair of variables (w_i, z_i) called the nonbasic pair. Lemke's algorithm operates by selecting an element of the nonbasic pair and a basic variable (*pair selection*) and then interchanging the chosen pair using pivoting (*pair interchange*). Pair interchange retains the complementarity and produces a new solution to the modified problem. If this new solution is such that $z_0 = 0$, then it is also a solution to the original problem. Otherwise, the process of pair selection and pair interchange is repeated. The algorithm may sometimes fail to terminate at a solution. Now we describe the actual algorithm informally using the following example from[70]:

$$w = \begin{bmatrix} 2 & -2 \\ -1 & 4 \end{bmatrix} z + \begin{bmatrix} -1 \\ -4 \end{bmatrix}$$

$$w \geq 0, z \geq 0, w^T z = 0$$

First, form a tableau that consists of three sections. The first section (column) lists the initial set of basic variables. The second section has $2n+1$ columns corresponding to all the variables in w, z, and z_0. Initially, it lists the matrix $[I - M - e]$. The last section (column) describes q. The tableau for our example is shown below:

Basic set	w_1	w_2	z_1	z_2	z_0	q
w_1	1	0	-2	2	-1	-1
w_2	0	1	1	-4	-1	-4

Next, a row of q that has the most negative element is selected. Let it be row s. In our example, it is row w_2. In general, if there is no negative element in q, a trivial solution to the original problem is obtained.

Then, elementary row operations are performed to convert the z_0 column to an identity column with its s-th element equal to 1. This is called *pivoting*. It is equivalent to interchanging the nonbasic variable z_0 with the basic variable represented by the row w_s (w_2 in the example). The resulting tableau with the new set of basic variables is shown below.

Basic set	w_1	w_2	z_1	z_2	z_0	q
w_1	1	-1	-3	6	0	3
z_0	0	-1	-1	4	1	4

Thus far, we have obtained a solution to the modified problem. For our example, we have $w_1 = 3$, $z_0 = 4$, $w_2=z_2=z_1=0$. Since z_0 is in the basic set, there exists a pair of complementary variables in the nonbasic set. In our example, $w_2=z_2=0$.

In Lemke's algorithm, the pair to be interchanged is selected using the following rule:

- *Column selection*: Choose the complement of the basic variable that has just been switched to the nonbasic set to become a new basic variable. In our example, the basic variable that has just been switched to nonbasic is w_2. Thus, z_2 is chosen.
- *Row selection*: Perform the so-called "minimum ratio test" to determine which basic variable (row) is to be replaced. Consider the selected column, divide each nonnegative element by its corresponding element in column q, and then choose the row with the smallest ratio. In our example, the ratios computed are shown in the last column of the tableau below. Row w_1 is selected.

This is equivalent to identifying an element in the tableau to pivot. In our example, it is the element at row w_1 and column z_2.

Basic set	w_1	w_2	z_1	z_2	z_0	q	ratio
w_1	1	-1	-3	$\underline{6}$	0	3	$\dfrac{3}{6}$
z_0	0	-1	-1	4	1	4	$\dfrac{4}{4}$

Pair interchange is equivalent to using elementary row operations to pivot the chosen element. For our example, the new tableau after pivoting (where w_1 and z_2 are interchanged) is shown below.

Basic set	w_1	w_2	z_1	z_2	z_0	q	ratio
z_2	$\frac{1}{6}$	$-\frac{1}{6}$	$-\frac{1}{2}$	1	0	$\frac{1}{2}$	---
z_0	$-\frac{2}{3}$	$-\frac{1}{3}$	1	0	1	2	2

Note that the new basic set is $\{z_2, z_0\}$. Since z_0 is still in the basic set, the process of pair selection and pair interchange is continued. In our example, the last basic variable interchanged is w_1. So the column z_1 is chosen. The ratios computed for nonnegative elements in column z_1 are also listed in the last column of the tableau above. The minimum ratio is achieved by row z_0. Thus row z_0 is chosen. Pivoting on the element at row z_0 and column z_0 yields the following new tableau:

Basic set	w_1	w_2	z_1	z_2	z_0	q
z_2	$-\frac{1}{6}$	$-\frac{1}{3}$	0	1	$\frac{1}{2}$	$\frac{3}{2}$
z_1	$-\frac{2}{3}$	$-\frac{1}{3}$	1	0	1	2

Since z_0 is in the nonbasic set, a solution to the original problem has been obtained. The resulting solution is $(w_1, w_2, z_1, z_2) = (0, 0, 2, 3/2)$.

3.3.2 Fault Simulation via One-Step Relaxation

DC fault simulation by one-step relaxation[74] is inspired by the very success of the simple stuck-at-fault models used in digital fault-driven testing. Stuck-at faults are first-order abstractions of the failure mechanisms in digital integrated circuits. They are simple enough to be amenable to efficient fault simulation and test generation. Nevertheless, the most fundamental reason why stuck-at fault models are so successful and have enjoyed industry use is that test vectors computed using stuck-at fault models tend to work well in practice even when the actual failure mechanisms may not be adequately abstracted by stuck-at fault models. Therefore, what is really important for fault simulation and test generation IS NOT the exact absolute modeling of failure mechanisms. Rather, the modeling of failure mechanisms need only offer relative accuracy, in that it can correctly predict the behavior of a faulty circuit.

For fault-driven analog testing, a high degree of absolute accuracy in nonlinear DC-fault simulation is difficult to achieve, and is in fact not really needed. As opposed to fault-driven digital testing, analog testing has to consider tolerance ranges due to the application-environment and measurement noise. For example, the simulated voltage value at an output terminal is 5.5 V under nominal conditions and 5.1 V in the presence of a given fault. Suppose that the measurement noise is 0.7 V. Then, if the measured voltage value at the output terminal is 5.4 V, we really cannot be certain whether the circuit is fault-free or faulty. This is referred to as *ambiguity*[29]. Thus, even if the calculated value of the faulty output is 5.5 (an 8% error relative to the actual faulty output value), no harm is done! It yields the exact same fault-coverage and test-generation results. Second, the exact faulty behavior of an analog circuit depends crucially on the precise modeling of a fault. However, for analog circuits, precise fault modeling is difficult.

Even with exact fault models, numerical noise incurred during the numerical simulation process may lead to inaccurate solutions, i.e., approximate solutions. Even worse, nonconvergence may occur as a consequence of pursuing exact fault simulation[80]. The one-step relaxation approach avoids these numerical problems while being capable of evaluating the DC-fault coverage for a given DC test with almost the same accuracy as exact fault simulation, but with significantly less CPU time.

3.3.2.1   Circuit Interpretation

In general, DC simulation amounts to solving a system of nonlinear equations written in the following form:

$$f(x) = 0 \tag{3.14}$$

where x is the circuit variable vector (node voltage or branch current), and f represents the system of nonlinear functions.

The Newton-Raphson algorithm exploited in SPICE-like simulators for solving Eqn. 3.14 is an iterative process. It starts from an initial point $x^{(0)}$, and then iterates until the difference between $x^{(k)}$ and $x^{(k+1)}$ satisfies certain convergence criteria. More precisely, it consists of the following steps:

- Guess: $x^{(0)}$
- Solve: $J(x^{(k)})(x^{(k+1)} - x^{(k)}) = -f(x^{(k)})$ for $k=0,1,2,...$ until the convergence criteria is met,

where $J(x^{(k)})$ is called the *Jacobian* matrix of $f(x^{(k)})$ and is calculated as:

$$J(x^{(k)}) = \frac{\partial f(x^{(k)})}{\partial x^{(k)}} \tag{3.15}$$

Suppose that the DC simulation for the good circuit (without faults) has been performed. Let x_g denote the solution of the good circuit. The process of *one-step relaxation* of DC fault simulation is to carry the Newton-Raphson algorithm for *only one step* for the faulty circuit using the solution of the good circuit as the starting point. That is, the process solves the following set of linear equations:

$$J_f(x_g)(x_f^{(1)} - x_g) = -f_f(x_g) \qquad (3.16)$$

where $J_f(x_g)$ and $f_f(x_g)$ are, respectively, the Jacobian matrix and function vector $f_f(x)$ calculated at the point x_g, and $x_f^{(1)}$ represents the vector to be solved. Then, the $x_f^{(1)}$ obtained is used as an approximate of the accurate solution of the faulty circuit. That would be the solution obtained by continuing to perform the Newton-Raphson iteration until it converges.

One-step relaxation has the following very simple circuit interpretation: Take the linearized circuit at the solution point of the good circuit as the "good-circuit" model, and then consider fault simulation of the "good" circuit. The fault models here are linearized models of the actual faulty circuits, at the solution point of the good circuit. DC fault simulation of nonlinear analog circuits using one-step relaxation is equivalent to fault simulation of linear circuits.

3.3.2.2   Householder's Formula

In fact, the solution $x_f^{(1)}$ can be computed very efficiently without solving Eqn. 3.16. This is done by exploiting the fact that the Jacobian matrix (also called the circuit matrix) for a faulty circuit usually differs only slightly from that of the good circuit. Therefore, Householder's matrix updating formula can be used.

3.3.2.3   Experimental Validation

A set of experiments with 29 MCNC benchmark circuits have confirmed the validity of approximate fault simulation by one-step relaxation. For a majority of the circuits, one-step relaxation yields the same fault coverage as does exact fault simulation. For those circuits for which fault coverage with the two methods is different, the difference is only a few percent. This observation is true for DC testing based on voltage measurement, as well as supply current monitoring. Since one-step relaxation can be performed rapidly with Householder's formula, this approach requires virtually no CPU-time overhead for fault simulation.

3.3.3   Simulation by Fault Ordering

The simulation continuation approach was proposed recently to improve the convergence problem associated with, and to reduce the number of iterations required by, the Newton-Raphson algorithm for exact DC fault simulation of nonlinear analog circuits[75]. It is based on the fact that whether the Newton-Raphson algorithm can converge, and how fast it converges if it does, depends crucially on the initial point used. The idea is to *order* faults in such a way that the results of one simulation can serve as a "good" starting point for the simulation of another fault, i.e., *fault simulation continuation*. Efficient one-step faulty response prediction derived from Householder's inverse matrix formula can be used to predict the "closeness" of the responses of the two faults. A greedy fault-ordering strategy has been developed to construct a near-optimum simulation continuation order. This approach is very effective for fault simulation with many parametric faults and for DC power supply testing where DC supply voltages are varied[67]. Experimental results with a set of MCNC difficult-to-simulate circuits demonstrated that for a mixture of injected hard faults and parametric faults, this approach can reduce the total number of iterations by a factor of 5-10. It is especially effective when a lot of faults, all of which are difficult to detect, yield similar behaviors.

3.3.3.1   Why Fault Ordering?

In general, DC simulation amounts to solving a system of nonlinear equations. The *Newton-Raphson* algorithm for solving Eqn. 3.14 is an iterative process that starts from an initial point x^0, and then iterates until the difference between x^k and $x^{(k+1)}$ satisfies certain convergence criteria. How close the initial point x^0 is to the final solution significantly affects the ability of the Newton-Raphson to converge, and determines how fast it converges[79]. However, predicting a good initial point in general is an unsolved problem, SPICE simply sets x^0 to be 0[36]. A very important technique that has been exploited for improving the convergence property for some difficult-to-converge circuits is *homotopy/simulation continuation*[26,49,82]. The basic idea is to construct a slightly different circuit whose solution can be a *good* initial point for the simulation of the original circuit, and furthermore, modify the circuit such that the simulation of the modified circuit can easily achieve convergence. Two simple schemes known as *Gmin stepping* and *source stepping* are implemented in SPICE[26,36], while more complicated schemes involve the design of traces of circuits, known as homotopy[49,82].

This idea can be exploited to accelerate DC fault simulation by ordering a given list of faults in such a way that the result from one fault simulation reduces the total number of iterations needed for simulating the next fault in the list. In the case of difficult-to-converge faults, more "pseudo" faults can be constructed and embedded into the simulation sequence to help convergence. This reduces the total number of iterations over all the faults.

The idea is especially attractive for parametric fault simulation, where the solution of the faulty circuit may be very close to that of the good circuit. For catastrophic faults, their solutions may differ significantly from each other and/or from that of the good circuit. Therefore, 0 or the solution of the good circuit may be the best initial point. This leads to the simulation trace as illustrated in Figure 3-6, where the solid circles represent solutions of faulty circuits that have to be simulated, the two shadowed triangle points correspond to the good circuit solution and 0 respectively, and the unfilled circles correspond to some "pseudo" faulty circuits solutions added to help the convergence of the simulation of some faults (using SPICE Gmin and source stepping for example).

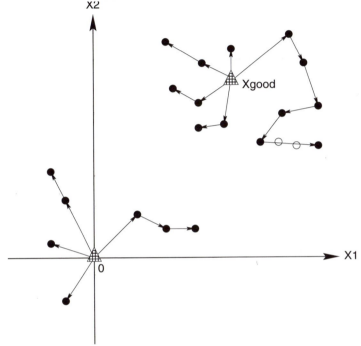

Figure 3-6 *Fault ordering via one-step relaxation.*

3.3.3.2 Fault Ordering via One-Step Relaxation

The problem of fault ordering, i.e., that of constructing the appropriate trace as illustrated in Figure 3-6, for simulation continuation can be solved by a computationally efficient yet effective heuristic. It consists of three components. First, all the faults are simulated using one-step Newton-Raphson iteration for all the faults with the solution of the good circuit as an initial point. The results are used as estimates of the actual solutions of faulty circuits. This step can be performed using Householder's formula. Second, a simple formula is used to "quantify" heuristically how close any two fault responses are. Finally, a simple greedy heuristic is used to order the faults. Given a nonzero solution vector x_i and a solution vector x_j, we define the *normalized absolute distance* as:

$$t_{ij} = \frac{1}{n} \sum_{k=1}^{n} \left| \frac{(x_i)_k - (x_j)_k}{(x_i)_k} \right| \tag{3.17}$$

Here $(x_i)_k$ and $(x_j)_k$ denote respectively the k-th component of the vectors x_i and x_j. The dimensions of the vectors x_i and x_j are n. A special case is the distance between x_{good} and x_j which is represented as t_j. The normalized absolute distance is a simple and yet effective measurement of the "closeness" of two faults. Note that the distance of any solution vector from 0 is always 1.

Now we can describe the adaptive and greedy fault-ordering strategy. First, one-step faulty response prediction is performed for all the given faults. The (normalized absolute) distance from each fault to the good circuit is then computed. Afterwards, the fault with the minimum distance is picked and simulated: If the distance is less than 1, then we simulate this fault using the solution of the good circuit. Otherwise, 0 is used. After we obtain the new simulated result, then we calculate the distances of remaining faults in the list, and identify the fault with the minimum distance. If this distance is less than the distance between the next fault in the list and the solution of the good circuit, and is also less than 1, we exchange the next fault with the fault with the minimum distance and simulate it. Otherwise, the next fault in the list is simulated. Whether 0, the solution of the current fault simulation, or the solution of the good circuit, is used for the next fault simulation depends on its distance. The details are described in Figure 3-7.

3.3.4   Handling Statistical Variations

To consider process variations, Monte Carlo random sampling has been used by several researchers in their recent work. However, the Monte Carlo approach is computationally expensive, and can only be used for small circuits with a few statistical parameters. To remedy this problem, Spinks and Bell proposed the use of Monte Carlo simulation to compute the worst-case boundaries for the good circuit, Then the band-fault assumption is used to compute the worst-case boundaries for faulty circuits[69]. This approach has been shown to work for some test circuits. However, in general, the good circuit and faulty circuits have different worst-case corners. We note also that Monte Carlo simulation always underestimates the response bounds and produces nonrobust test sets. A fault undetectable at a given test point may be claimed to be detectable by Monte Carlo simulation.

One-step relaxation-based fault simulation can be used in conjunction with Monte Carlo to approximate the fault bands of nonlinear analog circuits. Since one-step relaxation is rapid, the entire procedure can be very efficient. Using a linearized model for nonlinear analog circuits and then applying the interval-mathematics approach could be another option.

DC_Fault_Simulation_Continuation

1. $x_{good} \leftarrow$ *solution of the good circuit simulation;*
2. **for** *fault i in the fault list (i=1,...,m)*
3. $x_{fault\ i} \leftarrow$ *one step Newton-Raphson relaxation*;
4. $t_i \leftarrow$ *the distance from fault i to the good circuit*;
5. **if** $t_i > 1$, **then**
6. $x_{initial} \leftarrow 0$
7. **else**
8. $x_{initial} \leftarrow x_{good}$
9. **for** *fault i in the fault list (i=1,...,m)*
10. $x_{fault\ i} \leftarrow$ *simulation solution for fault i with* $x_{initial}$ *as the initial point*;
11. $x_{initial} \leftarrow x_{fault\ i}$;
12.   *calculate distance* t_{ij} *of fault j (j = i+1,...,m) to* $x_{fault\ i}$
13.   *let fault k have the maximum distance* t_{ik} *to* $x_{fault\ i}$
14.   **if** $t_{ik} > 1$ *and* $t_{ik} > t_{i+1}$
15.   *exchange fault i +1 and fault k*
16.   **if** $t_{ik} > t_{i+1}$ *and* $t_{i+1} < 1$
17. $x_{initial} \leftarrow x_{good}$
18.   **if** $t_{ik} > 1$ *and* $t_{i+1} > 1$
19. $x_{initial} \leftarrow 0$

Figure 3-7   *DC fault simulation algorithm based on simulation continuation and fault ordering.*

3.3.5   **Summary**

Nonlinear DC fault simulation is an important but difficult problem. The complementarity pivoting method developed in the early 1980s remains interesting. It is based on a mathematical concept totally different from the Newton-Raphson method found in SPICE-like simulators. This method does not suffer the same numerical problem for handling faults with extreme value changes and extreme nonlinearity in the circuit. One-step relaxation and simulation continuation are two methods developed very recently to improve the use of the Newton-Raphson algorithm for nonlinear DC fault simulation. They have been demonstrated to be effective and efficient, and are therefore ready for industry adoption into commercial fault simulators. Under the title of concurrent analog fault simulation, it has been proposed that the cache mechanism in a computer can be used to avoid the repeated evaluation of the Jacobian matrix for nonlinear devices, if they have the same terminal voltage/currents between simulations[85]. This technique has been exploited by several SPICE-like simulators, and is effective for the simulation of certain classes of circuits where latency is inherent. The class of circuits for which this technique is effective, and its actual effectiveness when applied, requires further investigation.

3.4   Fault Co-Simulation with Multiple Levels of Abstraction

We use the term *fault co-simulation* to refer to the simulation of a faculty circuit and system where portions are described at various abstraction levels and possibly simulated with different algorithms. Co-simulation is made possible by analog and mixed-signal hardware description languages and simulators supporting the languages. We note that hardware description languages, such as VHDL and Verilog, have been around for more than a decade. However, fault co-simulation is just a new area due mainly to the fact that faults can only be meaningfully modeled at the logic level or below (analog). Fault co-simulation not only enables system-level fault simulation, but also empowers traditional digital or analog fault simulation with a varied set of realistic faults models spanning several abstraction levels (stuck-at faults, bridge faults, etc.).

3.4.1   Mixed-Signal Simulators

Mixed-signal simulators can simultaneously simulate the analog portion of a design at the nonlinear differential equation level using Spice-like integration methods and the digital portion using the event-driven mechanism. Examples of such simulators include commercially available Saber[50] and Eldo[16], and an academic research tool IMacsim[52].

Two representative works done along this line are by Caunegre and Abraham at Siemens Automotive in France using Saber[5,6], and Sebeke, Teixeira, and Ohletz at the Universitat Hannover in Germany using Eldo[51]. Caunegre and Abraham described a procedure to build a fault simulator based on Saber for the fault simulation of mixed-signal systems. One of their contributions is that they show how digital-bridge simulation, which cannot be achieved with classical fault simulators, can be done nicely in the mixed-signal simulation environment. In their work, logic gates are replaced by mixed-signal components with digital inputs and analog outputs. A bridge fault is modeled by a local analog fault in the portion of resistive network that models the output stages of logic gates. Experiments on an adder and an A/D converter have demonstrated that mixed-signal fault simulation can achieve almost the same accuracy as transistor-level simulation but with 5 to 16 times speedup. It has also been observed that analog faults are time-expensive to simulate.

Sebeke, Teixeira, and Ohletz implemented a fault simulator called AnaFault based on the Eldo simulator and layout-realistic fault extraction. A fabricated integrated CMOS VCO was used to demonstrate the effectiveness of their simulator.

Both efforts observed that using these types of mixed-signal simulators, resistor models of short/open faults require less CPU time than the source models. It has also been observed that the choice of resistor values can have a significant impact on fault simulation results[51].

3.4.2   Incorporating Behavioral Models in Fault Simulation

Exploiting the co-existence of multiple levels of abstractions in a single system can speed up transistor-level fault simulation. This is achieved by first partitioning the transistor-level circuits into a list of subcircuit blocks and then modeling those blocks that do not have faults at the higher level (at the logic-level, or even a behavioral level) of abstraction. This approach has been implemented in several industrial environments based on commercial mixed-signal simulators.

Harvey, Richardson, Bruls, and Baker from Lancaster University, UK and Phillips Research Labs at Netherlands exploited this approach to evaluate the effectiveness of a number of test methods for a mixed-signal Phase Locked Loop (PLL) circuit. They partitioned the PLL into subcircuit blocks, and then added the target fault to a subcircuit that is described at the transistor level, while all other subcircuits are described at the behavioral level. Behavioral models are described using voltage-controlled voltage sources (VCVS) and voltage controlled current sources (VCCS). This allows the use of PSTAR, a Phillips SPICE-like circuit simulator. Nearly all the models are 10 to 36 times faster than the transistor-level description. This research is reviewed in more detail in Chapter 2.

3.4.3   Fault Macromodeling and Induced Behavioral Fault Modeling

One idea to further speed up transistor-level fault simulation is to model faults at the subcircuit block level. This is first done using the macromodeling concept widely used in SPICE-based circuit simulation (called fault macromodeling). Macromodeling refers to the process of deriving a compact model, using the circuit primitives available in a circuit simulator, for a transistor-level subcircuit block. This concept has been used by several researchers including Soma for sample and hold circuits and active filters[64,63], Meixner and Maly for a CMOS operational amplifier[33], and Pan, Cheng, and Gupta for Opamps[43].

There are three advantages to this approach. First, it uses less CPU time due to the use of macromodels. Second, only faults observable at the macro outputs are considered. These fault models are the most germane to manufacturing testing, and in functional testing. Finally, it leads to a smaller more manageable fault list, good for system-level fault testing, and diagnosis.

For historic reasons, macromodeling only permits models to use available primitive circuit components. This could be extremely tedious and complicated. Recent advances in analog and mixed-signal hardware description languages provide the ability to model the circuit behavior directly at the behavioral level using language constructs such as equations and if-then-else statements. This approach is called induced behavioral fault modeling in[58]. When the transistor-level fault list is extracted from inductive fault analysis, this approach does not suffer the same problem as digital behavioral fault modeling where behavioral faults are created as language permutations (no physical meanings)[2].

The behavioral fault modeling methodology assumes the existence of a behavioral/macro model for the circuit without faults. Then fault modeling consists of the following major steps (or a slight variation):

- *Inductive Fault Analysis*: Use inductive fault analysis to derive a set of realistic faults at the transistor level from actual layouts and process data.

- *Fault Screening and Grouping Using Performance Specification*: For each fault in the initial fault list, circuit simulations using an analog simulator are performed to see whether the fault manifests itself as a circuit malfunction. If the circuit shows no performance degradation during functional testing, then the fault is unobservable and is discarded. Other approaches such as sensitivities can be used. The faults that yield similar characteristics are grouped.

- *Fault Abstraction*: Introduce modification and extra elements into the fault-free model which would be used to deform its behavior in the manner characteristic of each of the groups identified in the previous step

- *Model Validation and Iteration*: Validate the model by simulating and measuring the difference between the obtained responses. When the difference is too large, the fault model is modified by repeating the steps above.

The lack of a procedure to extract these macromodels automatically is a bottleneck that remains. For those fault models which have the same structure (equations) as the good behavioral model, but only differ in parameters, stochastic optimization has been used[58]. Recently, a sensitivity-enhanced genetic algorithm has been demonstrated to be very effective[62].

3.4.4 Statistical Behavioral Modeling

Statistical behavioral models are behavioral models that incorporate statistical variations of parameters. For the purpose of functional verification, behavioral modeling and simulation focuses on the nominal behavior of the circuit. In the situation where the fluctuations of the circuit due to process and environment variations must be considered, then statistical behavioral modeling is important. The application that originally motivated statistical behavioral modeling was the need to determine the manufacturing yield[13]. Some work on systematic methods for statistical behavioral modeling were reported by Shao and Harjani[53], Koskinen and Cheung[27], and Styblinski and his students[46,71]. Initial efforts to exploit statistical behavioral modeling for mixed-signal testing include [7] for analog filters and Liu, Felt and Sangiovanni-Vincentelli for A/D converters[19].

Similar to behavioral modeling for nominal circuit performances, statistical behavioral modeling can be partitioned into two problems, depending on whether a parametric behavioral model is available or not. If a parametric behavioral model is given, then the problem is that of mapping the process and geometric device variations expected during manufacture onto the behavioral parameters. A full parameterized statistical behavioral model includes not only typical values, but also ranges of the mean, standard derivations and correlations of these parameters. In some cases, the mapping can be obtained directly from the performance values using analytical equations such as those in operational-amplifier macromodeling and switched-capacitor filters[7]. A general method was developed by Koskinen and Cheung to derive statistical models of analog and digital circuit cells at the behavioral level, which requires no prior assumption about the underlying analytic mapping[27]. It consist of the following steps:

- Calculate the nominal performance vector of the circuit and then map it to the nominal behavioral parameter vector for a given behavioral model. Note that the mapping is unknown, but the inverse $P=F(B)$ can be evaluated using behavioral simulation. This is done using the Downhill Simplex method such that the distance between P_0 and $F(B_0)$ is minimized in the least square sense. The issue is that of identifying a good initial point.
- Calculate the sensitivity matrix (change of B_0 in responses to changes of p_0) by perturbing the behavioral parameters around the nominal B_0 and resimulating the circuit.
- Derive the nominal, standard deviation and distributions, and correlations of the behavioral parameters using the sensitivity information: $B = B^0 + S\Delta P$
- Assume that the sensitivity matrix remains the same around the nominal regions of the P and B space, This can be extended with the Newton method to improve the accuracy of B.

In case behavioral models are not explicitly given, several methods have been proposed to construct analytic functions relating the behavioral parameters to a set of input parameters using the response surface methodology. These include modeling using radial basis functions[53] or using additive functions[19,71].

While statistical behavioral modeling is emerging as one of most important research areas for mixed-signal testing, several fundamental challenges remain. First, systematic methods and tools need to be developed. Second, some theoretical justification and issues need to be investigated. For example, what are the physical interpretations of distributions at the behavioral level. Third, more industry experimental and application demonstrations are needed. An issue that may affect the use of statistical behavioral modeling is that most of the time, detailed data are proprietary information, restricted to the silicon foundry. Testing is done at the silicon foundry, but behavioral modeling (as opposed to the traditional transistor-level netlist which can be extracted from layout) has to be done by designers. Therefore certain design flow/methodology issues need to resolved. Nevertheless, deep submicron mixed-signal design calls for a tight integration of the high-level design process with the manufacturing process. The arrival of analog and mixed-signal hardware description languages and standards has provided a basis for this new methodology.

3.4.5   Remarks on Hardware Description Languages

Mixed-signal hardware description languages are the foundation for fault co-simulation with multiple levels of abstraction. They also allow the integrated modeling of test environment and test equipment, and consequently enable the concurrent development of test plans in the design phase. A big challenge for adopting this methodology is the development of libraries of behavioral models and behavioral fault models, as well as the standardization of modeling conventions.

3.5 Concluding Remarks

In this chapter, we have reviewed various methods for efficient and effective fault simulation of analog and mixed-signal circuits. In particular, we examined methods for linear analog circuits, DC fault simulation methods, and system-level fault simulation exploiting partitioning and multi-level abstractions.

Analog and mixed-signal fault simulation is still an emerging and challenging research area. Technologies are far from mature to enable us to deal with the current state-of-the-art mixed-signal (including RF) chips and systems. We hope this chapter can provide a starting point for more future research.

References

1. *Semiconductor Industry Association*, Study Report, 1993.
2. J.A. Armstrong, F.S. Lam, and P.C. Ward, "Test generation and fault simulation for behavioral models," Chapter 5 in *Performance and Fault Modeling with VHDL*, J.M. Schoen (ed.), Prentice-Hall, 1995.
3. B. Ayari, N.B. Hamida, and B. Kaminska, "Automatic test vector generation for mixed-signal circuits," in *Proc. European Test Conference*, pp. 458-463, 1995.
4. J.W. Bandler and A.E. Salama, "Fault diagnosis of analog circuits," *Proc. of IEEE*, pp. 1279-1325, 1985.
5. P. Caunegre and C. Abraham, "Achieving simulation-based test program verification and fault simulation capabilities for mixed-signal systems," in *Proc. European Design and Test Conf.*, pp. 469-477, 1995.
6. P. Caunegre and C. Abraham, "Fault simulation for mixed-signal systems," *Journal of Electronic Testing: Theory and Applications (JETTA)*, vol. 8, no. 2, April 1996.
7. C.-Y. Chao, H.-J. Lin, and L. Milor, "Optimal testing of VLSI analog circuits," *IEEE Trans. Computer-Aided Design*, vol. 16, no. 1, pp. 58-77, Jan. 1997.
8. A. Chatterjee, B.C. Kim, and N. Nagi, "DC built-in self-test for linear analog circuits," *IEEE*

Design & Test of Computer, pp. 26-33, June 1996.

9. H.S.M. Chen and R. Saeks, "A search algorithm for the solution of the multifrequency fault diagnosis equations," *IEEE Trans. Circuits and Systems*, vol. 26, no. 7, pp. 589-594, July 1979.

10. L.O. Chua, *Introduction to Nonlinear Network Theory*, McGraw-Hill, New York, 1969.

11. G. Devarayanadurg and M. Soma, "Analytical fault modeling and static test generation for analog ICs," in *Proc. International Conference on Computer-Aided Design*, pp. 44-47, 1994.

12. G. Devarayanadurg and M. Soma, "Dynamic test signal design for analog ICs," in *Proc. International Conference on Computer-Aided Design*, pp.627-630, 1995.

13. S.W. Director, W. Maly, and A.J. Strojwas, *VLSI Design for Manufacturing: Yield Enhancement*, Kluwer Academic Publishers, 1990.

14. D. Domanchik, "New mixed-signal ICs demands sophisticated tests," *Test & Measurement World*, vol. 15, no. 6, pp. 24-30, May 1995.

15. P. Duhamel and J.C. Rault, "Automatic test generation techniques for analog circuits and systems: A review," *IEEE Trans. Circuits and Systems*, vol. 26, no. 7, pp. 411-440, 1979.

16. *Eldo Simulator User Manual*, AnaCAD/Mentor Graphics, Inc., 1990.

17. B.R. Epstein, M. Czigler, and S.R. Miller, "Fault detection and classification in linear integrated circuits: An application of discrimination analysis and hypothesis testing," *IEEE Trans. CAD*, vol. 12, no. 1, pp. 102-113, Jan. 1993.

18. B.C. Eves, "Linear complementarity and piecewise linear equations," *Homotopy Methods and Global Convergence*, B.C. Evaes, F.J. Gould, H.-O. Peitgen, and M.J. Todd, eds., New York: Plenum, pp. 79-80, 1983.

19. E. Felt and A.L. Sangiovanni-Vincentelli, "Testing of analog systems using behavioral models and optimal experimental design techniques," in *Proc. International Conference on Computer-Aided Design*, pp. 672-678, 1994.

20. P. Feldmann and R.W. Freund, "Reduced-order modeling of large linear subcricuits via a Block Lanczos Algorithm," in Proc. *IEEE/ACM 32rd DAC*, pp. 474-478, June 1995.

21. A. Grochowski, D. Bhattacharya, T.R. Viswanathan, and K. Laker, "Integrated circuit testing for quality assurance in manufacturing: history, current status, and future trends," *IEEE Trans. Circuits and Systems II: Analog and Digital Signal Processing*, vol. 44, no. 8, pp. 610-633, August 1997.

22. R. Harjani and B. Vinnakota, "Mixed signal test," Chapter 7.8 in *Technology for Multimedia*, Collection of Tutorials at *IEEE International Symposium on Circuits and Systems*, 1995.

23. C.W. Ho, A.E. Ruehli, and P.A. Brennan, "The modified nodal approach to network analysis," *IEEE Trans. Circuits and Systems*, vol. 22, pp. 504-509, June 1975.

24. A.S. Householder, "A survey of some closed methods for inverting matrices," *SIAM J. Appl. Math.*, vol. 5, pp. 155-169, 1957.

25. J.E. Jagodnik and M.S. Wolfson, "Systematic fault simulation in an analog circuit simulator," *IEEE Trans. Circuits and Systems*, vol. 26, no. 7, pp. 549-554, July 1979.

26. K.S. Kundert, *The Designer's Guide to SPICE & SPECTRE*, Kluwer Academic Publishers, 1995.

27. T. Koskinen and P.Y.K. Cheung, "Hierarchical tolerance analysis using statistical behavioral models," *IEEE Trans. CAD*, vol. 15, no. 5, pp. 506-516, May 1996.

28. C.E. Lemke, "Bimatrix equilibrium points and mathematical programming," *Manag. Sci.*, vol. 11, pp. 681-689, May 1965.

29. P.M. Lin and Y.S. Elcherif, "Analogue circuits fault dictionary - new approaches and implementation," *Journal of Circuit Theory and Applications*, vol. 12, pp. 149-172, 1985; Also in *Selected Papers on Analog Fault Diagnosis*, R.W. Liu, IEEE Press, 1987.

30. Ruey-Wen Liu, *Selected Papers on Analog Fault Diagnosis*, IEEE Press, 1987.

31. W.J. McCalla, *Fundamentals of Computer-Aided Circuit Simulation*, Kluwer Academic Publishers, 1988.

32. Matthew Mahoney, *Tutorial: DSP-Based Testing of Analog and Mixed-Signal Circuits*, IEEE Computer Society Press, 1985.

33. A. Meixner and W. Maly, "Fault modeling for the testing of mixed integrated circuits," in Proc. *IEEE International Test Conf.*, pp. 564-572, 1991.

34. L. Milor and A.L. Sangiovanni-Vincentelli, "Minimizing production test time to detect faults in analog circuits," *IEEE Trans. Computer-Aided Design*, vol. 13, no. 6, pp. 796-813, June 1994.

35. W.W. Mao, Y.S. Lu, R.K. Gulati, R. Dandapani and D.K. Goel, "Test generation for linear analog circuits," in *Proc. IEEE Custom Integrated Circuits Conference*, pp. 521-524, 1995.

36. L.W. Nagel, *SPICE2: A Computer Program to Simulate Semiconductor Circuits*, ERL Memorandum M520, Ph.D. Dissertation, Department of Electrical and Computer Engineering, University of California, Berkeley, CA, May 1975.

37. N. Nagi, A. Chatterjee, A. Balivada and J.A. Abraham, "Fault-based automatic test generator for linear analog circuits," in *Proc. IEEE International Conference on Computer-Aided Design*, pp. 88-91, 1993.

38. N. Nagi, A. Chatterjee, and J.A. Abraham, "DRAFTS: discretized analog circuit fault simulator," pp. 509-514 in *IEEE/ACM Design Automation Conf.*, 1993.

39. A.R. Newton, *Semiconductor Industry in Year 2010*, Report to DARPA, Oct. 1997.

40. A.R. Newton and A.L. Sangiovanni-Vincentelli, "Relaxation-based electrical simulation," *IEEE Trans. Computer-Aided Design*, vol. 3, no. 4, pp. 308-331, Oct. 1984.

41. M.J. Ohletz, "Realistic faults mapping scheme for the fault simulation of integrated analogue CMOS circuits," in *Int. Test Conf.*, pp. 776-785, 1996.

42. A. Pahwa and R.A. Rohrer, "Band-faults: Efficient approximations to fault bands for the simulation before fault diagnosis of linear circuits," *IEEE Trans. Circuits and Systems*, vol. 29, pp. 81-88, 1982.

43. C.-Y. Pan, K.-T. Cheng, and S. Gupta, "A comprehensive fault macromodel for Opamps," in *IEEE/ACM ICCAD'94*, pp. 344-348, 1994.

44. L.T. Pillage and R.A. Rohrer, "AWE: Asymptotic waveform estimation," *IEEE Trans. Computer-Aided Design*, vol. 9, pp. 352-266, 1990.

45. C.T. Pynn, "Combine test steps to reduce costs," *Test & Measurement World*, vol. 15, no. 6, pp. 39-46, May 1995.

46. M. Qu and M.A. Styblinski, "Statistical characterization and modeling of analog functional blocks," *Proc. IEEE Int. Symp. Circuits and Systems*, London, pp. 1.121-1.124, May-June 1994.

47. R. Ramadoss and M.L. Bushnell, "Test generation for mixed-signal devices using signal-flow graphs," in *Proc. International Conf. on VLSI Design*, Bangalore, India, 1996.

48. M. Renovell, G. Cambon, and D. Auvergne, "FSPICE: A tool for fault modeling in MOS circuits," *Integration: The VLSI Journal*, vol. 3, pp. 245-255, 1985.

49. J.S. Roychowdhury and R.C. Melville, "Homotopy techniques for obtaining a DC solution of large-scale MOS circuits," in *Proc. IEEE/ACM Design Automation Conf.*, pp. 286-291, 1996.

50. *Saber Simulator User Manual*, Analogy, Inc., Beaverton, OR, 1994.

51. C. Sebeke, J.P. Teixeira, and M.J. Ohletz, "Automatic fault extraction and simulation of layout realistic faults for integrated analogue circuits," in *Proc. European Design & Test Conf.*, pp. 464-468, 1995.

52. R.A. Saleh and A.R. Newton, *Mixed-Mode Simulation*, Kluwer Academic, Norwell, MA, 1990.

53. J. Shao and R. Harjani, "Macromodeling of analog circuits for hierarchical circuit design," in *Proc. Int. Conf. Computer-Aided Design*, pp. 656-663, 1994.

54. C.-J. Shi and K. Zhang, "A robust approach to timing verification," in *IEEE International Conf. Computer-Aided Design*, pp. 56-59, Nov. 1987.

55. C.-J. Shi and K. Zhang, "Tree relaxation: a new iterative solution method for linear equations," in

IEEE International Symp. Circuits and Systems, pp. 2355-2359, June 1988.

56. C.-J. Shi, "Analysis, sensitivity and Macromodeling of the Elmore delay in linear networks for performance-driven VLSI design," *International Journal of Electronics*, vol. 75, no. 3, pp. 467-484, Sept. 1993.

57. C.-J. Shi, "Entity overloading for mixed-signal abstraction in VHDL," in *Proc. European Design Automation Conf. (EuroDAC/EuroVHDL'96)*, Geneva, Switzerland, pp. 562-567, Sept. 16-20, 1996.

58. C.-J. Shi and N.J. Godambe, "Behavioral fault modeling and simulation of phase-locked loops using a VHDL-A like language," *Proc. IEEE International ASIC Conf. and Exhibit*, pp. 245-250, 1996.

59. C.-J. Shi and M. Tian, "Simulation and sensitivity of linear analog circuits under parameter variations by robust interval analysis," to appear in *ACM Transactions on Design Automation of Electronic Systems*, 1998.

60. C.-J. Shi and M. Tian, "Automated test generation for linear(ized) analog circuits under parameter variations," to appear in *Proc. ASP-DAC'98*, pp. 501-506, Feb. 1998.

61. C.-J. Shi and A. Vachoux, "VHDL-A design objectives and rationale," *Current Issues in Electronic Modeling*, Kluwer Academic Publishers, vol. 2, pp. 1-30, 1995.

62. C.-J. Shi, Y. Ye and X. Tan, "Behavioral model optimization via sensitivity-enhanced genetic search," *IEEE/VIUF International Workshop on Behavioral Modeling and Simulation*, Washington DC, pp. 17-24, Oct. 1997.

63. M. Soma, "A design-for-test methodology for active analog filters," in *Proc. International Test Conf.*, pp. 183-192, 1990.

64. M. Soma, "An experimental approach to analog fault models," pp. 13.6.1.-13.6.4 in *Proc. CICC'91* 1991.

65. M. Soma, "Challenges in analog and mixed-signal fault models," *IEEE Circuits and Devices*, pp. 16-19, 1996.

66. M. Soma, "Automatic test generation algorithms for analogue circuits," pp. 366-373 in *IEE Proc. Circuits Devices, Systems*, vol. 143, no. 6, Dec. 1996.

67. S.S. Somayajula, E.Sanchez-Sinencio and J.P. deGyvez, "Analog fault diagnosis based on ramping power supply current signature clusters," *IEEE Trans. Circuits and Systems Part II*, pp. 703-712, 1996.

68. T.M. Souders and G.N. Stenbakken, "Modeling and test point selection for data converter testing," in *Proc. International Test Conf.*, pp. 813-817, 1985.

69. S.J. Spinks and I.M. Bell, "A comparison of relative accuracy of fault coverage analysis techniques based on analogue fault simulation," in *Proc. 2nd IEEE Int. Mixed-Signal Testing Workshop*, pp. 17-22, 1996.

70. S.N. Stevens and P.-M. Lin, "Analysis of piecewise-linear resistive networks using complementary pivot theory," *IEEE Trans. Circuits and Systems*, vol. 28, no. 5, pp. 429-441, May 1981.

71. J.F. Swidzinski and M.A. Styblinski, "Statistical behavioral modeling and simulation: concepts and techniques," in *Proc. IEEE/VIUF International Workshop on Behavioral Modeling and Simulation*, Washington DC, pp. 41-48, Oct. 20-21, 1997.

72. G.C. Temes, "Efficient methods of fault simulation," in *Proc. 20th Midwest Symp. Circuits Syst.*, Texas Tech Univ., pp. 191-194, August 1977.

73. W. Tian and C.-J. Shi, "Rapid frequency-domain analog fault simulation under parameter tolerances," in *Proc. 34th IEEE/ACM Design Automation Conference*, pp. 275-280, 1997.

74. W. Tian and C.-J. Shi, "Nonlinear DC-fault simulation by one-step relaxation-linear circuit models are sufficient for nonlinear DC-fault simulation," in *Proc. IEEE VLSI Test Symposium*, 1998.

75. W. Tian and C.-J. Shi, "Efficient DC fault simulation of nonlinear analog circuits," to appear in

Prof. Design, Automation and Test in Europe Conference (DATE'98), Paris, France, Feb. 23-26, 1998.

76.   S.J. Tsai, "Test vector generation for linear analog devices," in *Proc. International Test Conf.*, pp. 592-597, 1991.

77.   L. Vandenberghe, B.L. De Moor, and J. Vandewalle, "The generalized linear complementarity problem applied to the complete analysis of resistive piecewise-linear circuits," *IEEE Trans. Circuits and Systems*, vol. 36, no. 11, pp. 1382-1391, Nov. 1989.

78.   P. Variyam and A. Chatterjee, "FLYER: fast fault simulation of linear analog circuits using polynomial waveform and perturbed state representation," in *Proc. International Conference on VLSI Design*, India, pp. 408-412, Jan. 1997.

79.   J. Vlach and K. Singhal, *Computer Methods for Circuits Analysis and Design,* Van Nostrand Reinhold Press, 1994.

80.   R.S. Vogelsong, 'Trade-offs in analog behavioral model development: managing accuracy and efficiency," in *Proc. IEEE/VIUF International Workshop on Behavioral Modeling and Simulation, Washington D.C.*, pp. 33-40, Oct. 1997.

81.   H. Walker and S.W. Director, "VLASIC: A catastrophic fault yield simulator for integrated circuits," *IEEE Trans. CAS*, vol. 5, no. 4, pp. 541-556, Oct. 1986.

82.   D.M. Wolf and S.R. Sanders, "Multiparameter homotopy methods for finding DC operating points of nonlinear circuits," *IEEE Trans. Circuits and Systems Part I*, pp. 824-838, 1996.

83.   H. Xue, C.N. Di, and J.A.G. Jess, "A net-oriented method for realistic fault analysis," in *Proc. IEEE Int. Conf. on Computer Aided Design*, pp. 78-84, Nov. 1993.

84.   P. Yang, *An investigation of ordering, tearing, and latency algorithms for the time-domain simulation of large circuits*, Ph.D. Dissertation, University of Illinois at Urbana-Champaign, 1980.

85.   W. Zwolinski, A.D. Brown and C.D. Chalk, "Concurrent analogue fault simulation," in *Proc. 3rd IEEE Int. Mixed-Signal Testing Workshop*, pp. 824-838, 1997.

Automatic Test Generation Algorithms

Mani Soma

4.1 Introduction

Automatic test pattern generation (ATPG) algorithms for analog circuits, speaking strictly from the viewpoint of ATPG for digital circuits and systems, do not exist. There is no algorithm similar to PODEM, which processes a netlist based on an accepted universal fault model, and generates a set of test waveforms and a measure of the quality, or fault coverage, of this test set. Attempts have been made, especially in the past five years, to develop such an algorithm and have met with different degrees of success. This chapter reviews some fundamental issues in analog test generation to set the stage for a more in-depth description of these recent algorithms: their basic ideas, their diversity, their applicability to specific circuit classes, their limitations, and possible improvements. Several new techniques, not yet widely verified, are also included to provide an indication of the future direction of analog ATPG.

4.2 Fundamental Issues in Analog ATPG

Considering the expertise already existing in digital system ATPG, the relatively slow pace in analog ATPG is somewhat surprising but not unexpected. The universal stuck-at fault model, first described in the late 1950s, did not really give rise to powerful test generation algorithms until the late1970s, when the system test complexity issues drove digital test researchers to rely on structural tests and fault models rather than on functional specifications. The complexity in analog test does not have the same basis as that in digital test: usually the number of inputs, outputs, and devices is small (excluding relatively new circuits such as 22-bit converters), and the exponential explosion 2^N, where N is the number of inputs, is not pertinent. The continuous-time, continuous-value waveforms and the dependency of circuit performance on topologies and component values are some of the fundamental reasons distinguishing analog and digital ATPG. Other reasons and issues are reviewed in details in this section.

4.2.1   Structural Test Versus Functional Test

Analog test traditionally focuses on the verification of circuit performance: a set of specifications, with tolerance windows, needs to be tested in both prototype and manufacturing test. Functional test, or equivalently parametric test, thus is the norm and given the relatively small number of circuit classes still designed using analog techniques, the need for automatic generation of functional test is not urgent. For each class of circuits, there are already accepted functional test sets for full prototype test and a smaller subset for production test. It is interesting to note that the quality and the cost of these functional tests have not been an issue until recently, when the integration of analog and digital circuits onto the same chip prompts a revisit to the accepted functional test paradigm. With respect to test quality, there has been anecdotal industrial evidence of circuits passing these tests yet failing in system application, and recent works, e.g. [1], show that some analog circuits with catastrophic faults do pass a conventional manufacturing test set. The cost of analog functional test, becoming an ever larger part of the overall system cost [2], is also a major impetus for alternative test generation techniques.

Structural test provides one of these alternatives. Structural fault models have been published and are generally divided into two classes: catastrophic faults and parametric faults. The catastrophic fault class consists of digital-like faults: stuck-at-0, stuck-at-1, short or bridging fault, and open fault. The parametric fault class consists of faults that affect component values, which ultimately change the overall circuit performance. Viewed in this manner, the parametric fault class is related to functional test rather than to structural test. Algorithms have been developed to generate tests based on either or both fault models, and will be described below.

4.2.2   Path Sensitization

The success of digital ATPG may be partially attributed to the concept of path sensitization: a path to excite a fault at a specific node and to propagate the fault effect to an observable output may be sensitized by manipulating input values to establish the path while simultaneously disabling other paths that interfere with the propagation of desired signals. The concept of path sensitization has not been really applied to the analog ATPG problem: given a fault, tests are generated to detect output variation due to the fault but there is no effort to select and sensitize paths. Since most analog tests are still generated for macro-level circuits (filters, converters, amplifiers, etc.), signal propagation to an observable output in a larger system has not been an issue, at least until the macro is embedded in a digital system. Several test generation algorithms explicitly relying on path sensitization and signal propagation need built-in design-for-test (DFT) circuits. These algorithms, which will be described below, are much better structured and defined but have found only limited use so far due to designers' reluctance to add DFT to their analog designs.

4.2.3   Measurement Impact on Test Generation

A digital test creates a difference at an output depending on whether the circuit is fault-free or faulty. This difference, between logic 0 and logic 1, does not impose any problem in measurement. In analog test, however, the success of a test and the detectability of a fault depend strongly on the output measurement. If a test gives rise to a difference within the measurement error of the test set up or the instrumentation, the test actually is not usable since in an experimental setting, the fault effect is masked by this error and a bad circuit is indistinguishable from a good circuit. This is an extremely important criterion in analog test generation and determines if an algorithm is experimentally applicable, as opposed to its usefulness in simulation studies. Simulated results are sufficient in digital ATPG but not so in analog ATPG.

4.2.4   Simulation Impact on Test Generation

Simulation tools for analog circuits, while remarkably improved compared to the early version of SPICE, are still unable to simulate large circuits efficiently and accurately. Process variations, temperature requirements, and device model issues all contribute to the uncertain accuracy of the simulated results, especially in high-performance circuits. Hierarchical simulators, e.g., Spectre HDL, AHDL, Saber, etc. seek to accelerate the simulation process by providing macromodel capability but in most instances, trade accuracy for speed. Test generation algorithms frequently use transfer functions to model good linear circuits, and depending on the fault models, rely on simulators to predict fault detection. If a fault induces totally abnormal operations (e.g., cases of catastrophic faults), fault simulation has to be performed at the circuit level with fault insertion done manually. If a fault induces performance variations (e.g., cases of parametric faults), hierarchical simulations are still possible, especially if the parameters of transfer functions are perturbed. The most notable fault simulation problem involves the open fault. Several techniques have been proposed [3] ranging from using a large resistor to modeling the fault with random voltage sources but none has gained a wide acceptance.

Fault simulation is a topic in itself and a critical component of test generation, and the lack of a good efficient analog circuit simulator compounds the problems in analog fault simulation. Many fault simulation algorithms have been published and will be mentioned briefly below in conjunction with the analog ATPG algorithms which employ them.

It should be noted at this point that there exist a large number of papers published under the topic of analog ATPG, yet they, strictly speaking, do not describe algorithms to generate tests. These papers focus on aspects such as:

1. post-test analysis assuming a given test input, e.g. [4], where the input frequency is known, [5] where the DC test vector is given and seems to be derived empirically, and many more.
2. tests are used without a well-defined algorithm to generate such tests, e.g. [6].
3. test signal generation on a tester, e.g. [7].
4. sensitivity as a means for fault-list generation, fault diagnosis, output analysis, and testability, without any description of how sensitivity is used in analog test generation, e.g. [8].

These algorithms will not be covered in the description below.

4.3 Test Generation Algorithms and Results

The diversity of analog test generation algorithms is best presented in a structure of four classes: functional test generation, structural test generation, test generation via automatic selection and ordering, and DFT-based test generation. Strictly speaking, the third class (test selection) is not ATPG but is included for completeness as a test reuse tool and also for its more immediate applicability in analog test generation.

4.3.1 Functional test generation algorithms

4.3.1.1 Empirical Functional Test Sets

Classical analog test sets are generated empirically using the circuit specifications and the waveforms regularly used in simulation: DC, sine, step, square, ramp, etc. Time-domain testing and frequency-domain testing are employed depending on which parameters to measure. While strictly speaking, this test generation technique does not belong to the class of analog ATPG (there is no software tool nor well-defined algorithm), these test sets are important since they are used widely in industry and any new test algorithm has to be calibrated against them to gain users' acceptance.

Empirical tests are usually macro-based and measure the macro performance parameters. For example, an analog filter is tested for 3-dB bandwidth, stop-band attenuation, pass-band ripple, stop-band ripple, delay, and large-signal slew rate. Each parameter has a tolerance window and the circuit passes a test if the measured parameter is within the window. There is no explicit fault model, be it structural or parametric, applied to the internal components of the circuit. Besides direct measurement, output data is also frequently analyzed by the Fast Fourier Transform (FFT) to extract the desired parameter(s) from the test(s). Nonlinear analog circuits, e.g., converters, are also tested using the same specification-based paradigm. The best references to empirical functional test waveforms and techniques are data sheets and application notes from the design and test industry.

While empirical test sets are the standard in analog testing, recent issues have raised questions about their quality, cost, and applicability. Empirical test sets tend to be quite large in prototype testing to verify the circuit performance in new designs. The test cost is thus on the high side and has been estimated at 30% of the overall manufacturing cost [2]. A smaller test set, used in manufacturing test once the design and yield are acceptable, is empirically selected from the full test set to reduce cost but, as pointed out above, some circuits containing faults may actually pass the manufacturing test set [1]. The test quality is not defined either with respect to pass/fail criteria or any fault coverage measure. The most critical challenge to the continued use of empirical tests is the availability of large analog circuits, e.g. 22-bit converters, and the increasing requirement for mixed-signal circuits. A standard histogram test applied to a 22-bit converter would require an extremely large number of tests and a very long test time to collect sufficient samples for valid statistical analysis. Mixed-signal circuits with embedded analog circuits and sensors make the analog circuits untestable based on this paradigm since there is no mechanism to apply the empirical test set to the circuit, unless explicit test points are provided. Path sensitization is one technique that would make macro-based empirical functional tests reusable at the system level; unfortunately this technique has not been well developed except in conjunction with DFT algorithms.

4.3.1.2   *Functional Analog ATPG*

The requirement to show test effectiveness using the empirical test sets as standard drives many developments in analog ATPG. Fault models are mixed: Some algorithms assume only catastrophic faults, some only parametric faults, and some include both. Many algorithms assume that a linear circuit, when faulty, remains linear, which is appropriate for the parametric fault model but questionable for the catastrophic fault model. Linear system theory provides the foundation for these techniques.

An interesting algorithm uses signal flow graphs (SFG) as a test generation tool by reverse simulation and sensitizing the graph edge containing the faulty component [9]. The algorithm builds an SFG of the good linear circuit, reverses the directions of all edges, uses symbolic edge weights to represent faulty components, and calculates the required input to detect a fault. This analog backtrace technique has been applied to some nonlinear analog blocks and can be interfaced directly with the standard digital backtrace for mixed-signal circuits.

Instead of SFG, transfer functions are used in several other algorithms dealing with parametric faults and in some cases, including catastrophic faults. For example, in [10] and [11], the transfer functions of good and faulty circuits are formulated, and a sinusoidal test is generated by searching for a frequency that maximizes the amplitude or phase difference between these two transfer functions. Golden-search technique and root-finding technique such as the false-position method are employed. Fault simulation is performed to drop the detected faults and the process is continued to generate tests for the remaining faults.

In [12], the ATPG problem is formulated as a quadratic programming problem and solved using heuristics. The transfer function is discretized and the parameters of the function are assumed to be fault-free or faulty. The input vector x is also discretized, and the difference between the good and faulty outputs is easily shown to be a homogeneous, positive, semidefinite, quadratic function. The maximization of this difference is thus a quadratic programming problem, with NP-hard complexity. Using a heuristics which eliminates terms of the form x_n^2 but leaves all other terms in the difference function, it is possible to solve the optimization problem to arrive at a test signal. Applied to a bandpass filter example, the algorithm generates a waveform that performs quite well, better than a step-response test or a simple sine wave test. Efforts to extend this algorithm to other classes of circuits have not met with significant success due to the optimization complexity and the lack of appropriate heuristics for other transfer functions or for nonlinear circuits.

Iterative procedures are used to generate multifrequency tests in [13]. The input voltage and tolerance threshold are computed in each loop and fault detection is performed using frequency-domain measures. If a fault is not detectable, a different input voltage is tried in the next loop. The algorithm goes into more details for fault diagnosis, which is not pertinent here. It has been applied to circuits with built-in self-test.

4.3.1.3   *Sensitivity-based ATPG*

Several ATPG algorithms rely on sensitivity of an output or an observable signal with respect to either a component value or a collection of process parameters. Sensitivity has been used extensively in analog fault diagnosis [14] and is readily applicable to test generation at first glance.

Classical sensitivity measures based on component tolerance are used in [15] as the first step in test generation. To cover both parametric faults and catastrophic faults, an incremental sensitivity is defined and used instead of the regular differential sensitivity. When an output parameter T changes due to a perturbation, large or small, of a component value x, the incremental sensitivity is defined as:

$$\rho_x^T = \frac{x}{T} \times \frac{\Delta T}{\Delta x} \qquad (4.1)$$

This sensitivity may be calculated directly via simulation or from the transfer function of the circuit. The test generation procedure computes the sensitivity matrix of the output parameters as functions of selected component values, and creates a connectivity matrix showing the topological relationship between observable signals and components and another connectivity matrix showing the relationship between the output parameters and the observable output nodes. A directed graph is constructed using the two connectivity matrices with branch weights equal to the computed sensitivities. A minimum cost flow problem is then formulated and solved by optimization tools. Note this algorithm does not generate a test waveform as such; it instead generates the list of output parameters that need to be measured to cover the assumed faults. The algorithm has been applied to amplifiers and also discusses possible extensions to cover the multiple-fault case.

Test waveforms, or more specifically test frequencies, are automatically generated by another sensitivity-based algorithm [16]. In this instance, after the sensitivity matrix between a set of output parameters and parametric component deviations has been computed, a search procedure, assisted by simulation, is used to find a set of frequencies of input signals such that the difference between good and faulty outputs is maximized. This ATPG assumes linear circuit behavior only, even in the presence of faults, and relies on a behavioral fault model that includes both parametric and catastrophic faults. The algorithm classifies a fault as undetectable if the maximum sensitivity of an output parameter to this faulty value is less than the detection threshold imposed by experimental measurements. Fault dropping is accomplished using a discretized analog fault simulator, DRAFTS, which performs analog behavioral simulation in the Z-domain.

Under the same assumption of linear faulty behavior, another ATPG procedure [17] uses complementary signals to generate tests in time domain rather than in frequency domain. The complementary signal α_n of linear time-invariant network with n poles $\{v_i, i = 1 \ldots n\}$ is a piece-wise constant waveform and has n-step amplitudes

$$\alpha_k = (-1)^k \exp \left\langle \sum_{i=1}^{k} v_i T \right\rangle \qquad (4.2)$$

where T is a time interval proportional to the inverse of the minimum pole value. For a good circuit, the response to this input vanishes at time $(n+1)T$. Parametric variations and faults will induce corresponding variations of the complementary signal amplitudes and the circuit response will fail to vanish, thus constituting a condition for fault detection. The generation of the complementary signal relies on an algorithm to estimate the system poles [18] and determines T by maximizing the sensitivity $\frac{\partial \alpha}{\partial v}$. The method has been applied to active filters and operational amplifiers.

Sensitivity of output parameters with respect to test frequencies and component values provides a basis for other test generation algorithms presented both as ATPG [19] and as techniques to increase analog testability [20]. The algorithm assumes sine wave inputs, and selects test nodes and frequencies to make faults observable. The sensitivities of output parameters are calculated in the frequency domain, and given a fault, a frequency is selected to maximize the output sensitivity with respect to this fault. In the multiple-fault case, sensitivity is used to select frequencies to distinguish different faults if these faults do not mask one another. The technique has been applied to various types of filters.

Several other sensitivity-based algorithms, e.g. [21], apply other optimization criteria or focus on specific classes of circuits to derive tests, also taking into account process parameter variations as the fundamental underlying cause of circuit failures.

One characteristic of sensitivity-based algorithms is that some type of cost function has to be optimized using sensitivity as a guide. There are various techniques to optimize depending on how the problem is formulated, but if the optimization algorithm approaches NP-complete, e.g. [22], suboptimal heuristics may always be used to arrive at a test set in a reasonable time. These commendable efforts to use sensitivity as the basis for ATPG are quite mature and have made incremental gain toward the solution of the problem. In fact, there exists an automatic sensitivity analysis tool, LIMSoft [23] [24], to provide a front-end input to analog test generation. Since sensitivity has a firm theoretical foundation, ATPG for parametric faults will likely be most successful using this paradigm.

4.3.2   Structural Test Generation Algorithms

The major distinguishing characteristic between structural test and functional test is the procedure to derive and model faults. Functional test frequently assumes that the components are faulty and generates the fault list using component value deviations or catastrophic faults. Structural test uses manufacturing defect statistics as the basis to generate the fault list, and the faults generated may be either catastrophic (open or short) or parametric (small change in component value). Once the fault set is established, the structural ATPG algorithms employ many search strategies to find a test set to detect the faults. These search strategies usually do not rely on sensitivity as such but still share the same characteristics of sensitivity-based algorithms in using statistical optimization or statistical simulation.

4.3.2.1   Resistive-based ATPG

Using resistors to model structural faults, a high-level reasoning algorithm [25] searches for a path from the fault site to the output and selects a path by building path segments based on sensitivity with respect to the given fault. The algorithm uses breadth-first search to find a path from the faulty branch to an output node. The sensitivity metric for each path is computed using current node voltages since the final voltage is unknown during the test generation process. The voltage vector is modified in direction of higher sensitivity and checked for convergence of solution. Once a path exists, a DC test is automatically generated in this manner and guaranteed to detect the fault. This interesting algorithm combines several interesting features in test generation: simple fault model, path search, and sensitivity. The use of a resistor to model faults, however, has some shortcomings: It is adequate for shunt faults but not valid for open faults or for many parametric faults.

4.3.2.2   Linear Programming Algorithms as ATPG

Using linearized fault macromodels, two related algorithms [26] [27] employ minimax linear programming techniques to generate DC and AC tests to detect structural faults. Starting with the defect-based catastrophic fault model, the first algorithm [26] considers process parameter variations and a discrete set of DC input values as the search space in which to maximize the difference between the good and faulty output. The search problem is formulated as a series of successive linear programming problems, solved iteratively to arrive at the best DC input as the test to detect the fault. The second algorithm [27] extends the first by considering AC test generation as well, using the same underlying fault model. Since the search space of all AC waveforms is infinite, the algorithm derives a suboptimal solution by assuming that the AC input is piecewise-linear, and generates the signal by computing the pair [t, x(t)] at each iteration step. The input waveform is thus constructed one point at a time and within each time interval; the minimax linear programming technique is used to maximize the output difference over the space of process parameter variations and the set of discretized voltage or current values to be set at the input. One interesting result of these algorithms is that a saturated ramp waveform has been shown to be quite effective in fault-based testing of analog circuits. This conclusion is also reached by [11] in a more heuristic manner: assuming a slope for the ramp waveform and iteratively adjusting the slope to maximize the number of faults detected. This algorithm generates only 2 ramp signals to detect 100% of the assumed faults.

4.3.2.3   IDD-based ATPG

Power-supply current tests have proven to be very efficient in detecting structural faults in digital circuits, and a recent paper [28] summarizes its possible applications to analog testing. As an example, [29] applies the technique as an ATPG tool to several analog circuits. This algorithm uses periodic AC input and computes the spectra of measured I_{DD} values as the basis for fault detection and for selection of best test waveform. The spectral computation avoids the problem of time-domain comparison, which is quite susceptible to errors, and takes into account the parametric fault model as well by using multivariate statistical analysis. An input current must provide a spectrum separable for fault-free and faulty circuits. The input selection uses hypothesis testing, given the spectra of currents resulting from the faults in the fault list. The algorithm does rely on Monte Carlo simulation to establish probabilities of detection, and thus the expected run time could be significant. Examples with operational amplifiers and filters show good fault coverage, comparable to voltage-based test generation.

Pulsing the power supply and measuring the current from the supply is the basis of another test generation algorithm [30], in which the time-domain characteristics as well as the frequency spectrum of the resulting pulsed current are analyzed for fault detection. It is interesting to observe that the time-domain waveforms permit better fault detection than the frequency spectrum in the example of a folded cascode operational amplifier. The ATPG algorithm in this case is quite simple, but the output analysis becomes important in measuring the test quality.

Changing voltages to induce fault-dependent currents provides a basis for several other analog ATPG algorithms. [31] uses a fixed power supply but varies the input signal either as a ramp or a pulse or a step, then measures the power supply current to detect faults. Applying to converter testing, this technique reports up to 87% fault coverage. Varying the power supply is also shown to amplify the effect of a fault and make the current test more effective especially with higher V_{DD}. [32] ramps the power supply and shows that the bias currents, which are functions of circuit topology and component values, can be used to detect faults. Each fault in the fault list is simulated to obtain a current signature, and Kohonen network is used to discriminate between faults. The time-domain responses in one case study show a clear difference between fault-free and faulty I_{DD} currents while V_{DD} is ramped.

4.3.2.4   Pseudo-random ATPG

Analog linear time-invariant circuits have been tested using a pseudo-random test generation technique similar to the digital LFSR. [33] assumes that such a circuit is embedded between two converters: a DAC at the input and an ADC at the output. A digital LFSR is used to generate the pseudo-random test patterns, which are converted by the DAC to the analog input to the circuit under test. The analog output is digitized by the ADC and a digital signature analysis is performed to check fault detection. The signatures are the mean of the output, the auto-correlation of the output, and the cross-correlation between the input and the output. The input/output correlation is guaranteed via the transfer function, at least in the analog domain.

Structural ATPG algorithms, while not as numerous as functional ATPG algorithms, show strong promise of more significant future developments. The fault model, derived from manufacturing statistics, automatically takes into account both catastrophic and parametric faults. The optimization procedures employed to generate or select tests are still very compute-intensive due to the large search space and the continuous nature of the output functions. Several possible improvements to this class of algorithms will require the integration of search strategies with path selection techniques, the decomposition of a larger circuit into smaller macros to limit the search space, and/or the use of digital waveforms to excite and propagate the faults irrespective of the circuit's conventional analog functions. The structural ATPG algorithms have been verified experimentally over a small set of examples, thus there is still question about their general validity.

4.3.3   ATPG Based on Automatic Test Selection Algorithms

Given the existence of the functional empirical test set, it is reasonable to expect that a better test set may be derived or selected from this set once a fault model is established. The criteria for test set selection are quite simple: either to cover 100% of the selected faults, or to reduce test cost by reducing the number of tests, to order tests to reduce test time, or to do all simultaneously. There is a large body of works on these algorithms, several representatives are now described.

A test selection procedure assuming catastrophic faults [34] combines the simulation of faulty circuits and sensitivity to select the optimal test set, taking into account process variation and measurement bounds. Given a fault list and a set of tests, the algorithm proceeds in two steps. In the first step, a test is selected using a simple heuristic which determines the measurement which deviates the most from the good measurement for all faults in the list. In the second step, the fault signature is computed and checked to see if it intersects with the good signatures. If the two sets of signatures are disjoint, the fault is classified as detected and dropped from the fault list. The new list is formed and a set of good signatures is computed before looping back to the first step again to select the next test. The algorithm uses a desired fault coverage or the exhaustion of the original test set as the termination condition.

The above algorithm has been subsequently improved in [35] [36]. In this work, the fault model is derived from statistical process parameter variations, then a test-ordering algorithm is executed based on permutation of the functional test set to generate a test set with reduced test time. Tests with very low fault coverage are eliminated, and the remaining tests are ordered in one of two approaches. The first approach orders the test set with those having high fault coverage first, then uses a permutation on this test set to select tests to reduce test time. The second approach optimally orders the tests using a directed graph and a test time predictor associated with each node in the graph. The algorithms have been applied to analog amplifiers and techniques to take into account measurement uncertainty and are discussed within the context proposed.

Another test selection algorithm relies on a vector-space decomposition of the output measurements into a set of linearly independent basis vector set [37] using QR decomposition. The cardinality of this basis is the minimal number of tests required. This approach requires a large number of measurements to provide a high statistical confidence in the test set, but it has proven to be effective in reducing test time. Graph-based techniques organize tests in a directed graph with branch weights as test costs [38]. Each node in the graph represents a possible state or a possible test set. Each branch represents a path from one state to another, and the branch weight is the average cost of performing the next test. Dijkstra's algorithm is then used to find the shortest path, which is equivalent to selecting the smallest test set to save time and cost. These algorithms have exponential complexity and have not been widely used.

For a specific class of analog systems which can be modeled efficiently as an additive combination of user-defined basis functions, [22] proposes the use of a statistical confidence interval to verify the system performance and to choose test points to make the confidence interval as tight as possible. Given an arbitrary basis of n functions spanning n-dimensional space, at least n tests must be applied in order to characterize the system. From an existing test set, the algorithm eliminates redundant basis vectors to form a set of maximally orthogonal vectors and uses the I-optimality to select best n tests in the sense that the tests chosen minimize the average prediction variance of the circuit response. Additional tests are generated in the same manner to detect more faults if necessary. Since this optimization algorithm is believed to be NP-complete, gradient descent techniques are employed in the search. The algorithm has been applied to bandpass filters and digital-to-analog converters.

Another test selection algorithm exploits the circuit sensitivity, given the process variations and performance bounds on output parameters. The algorithm uses characteristic observation inference [39] and proceeds as follows. With the given input test set, the algorithm performs a sensitivity analysis for all measurements given the statistical process and parameter variations at a nominal point. The sensitivity matrix is processed to select the most linearly independent and most sensitive tests, using the Householder algorithm. Double iterations to generate tests from initial test seeds [40], most likely provided by existing tests or test engineers, use gradient algorithms and take into account defect size statistics in the test generation process. Parallel optimization to process many faults at one time are attempted even though circuit simulation in the inner loop still requires a large amount of CPU time.

Automatic test selection algorithms have the advantage of using the existing functional tests as the basis and thus should be easily acceptable to analog designers and test engineers. The theoretical foundation from statistics provides numerous techniques to select tests depending on the optimizing criteria: minimal number of tests, maximal fault coverage, minimal test time, etc. Whether these tests themselves are optimal when compared to tests generated by other techniques are still open to questions and further study.

4.3.4   DFT-based Analog ATPG Algorithms

Once a design modification is permitted to improve testability, the spectrum of algorithms for analog ATPG becomes almost infinite. Since analog DFT methodologies have not been standardized, there are numerous DFT techniques and for each, there is an ATPG. The comparison in algorithm efficiency is thus extremely difficult since it needs to take into account performance impact of the proposed DFT technique: layout size, loading, test benefit, etc. For the sake of completeness, we will review very briefly several DFT-based analog ATPG algorithms in this section. The review in this section focuses on the generation of tests with DFT schemes, not on the specifics of the scheme itself. Chapter 4 contains a detailed review of the DFT techniques discussed in this section.

The first class of DFT strategies employs redundancy [41] [42] [43] [44], and thus it is expected that some form of comparison or voting is required to validate the fault-free performance of the original circuit. Redundancy takes either one of two forms: complete redundancy or duplication, or partial redundancy using programmable circuits to reduce layout impact of the DFT addition. While fault models are not explicitly discussed, the ATPG algorithms generate functional tests, and these techniques relate more to functional rather than structural testing.

The second class of DFT strategies relies on the concept of controllability and observability to generate tests. Circuits are reconfigured by adding in switches in well-defined locations [45] so signals may be propagated with no bandwidth limitations in the test mode, or in switched-capacitor or switched-current systems, re-timed [46] so that selected paths can be sensitized during test. The test generation algorithms are quite simple since they rely on the concept of path sensitization, as in the case of digital systems. Once a fault is selected, the fault excitation and propagation steps are identical to digital ATPG, and the detection depends on the accuracy of the output measurement and the parameter tolerance. The DFT and ATPG algorithms [45] have been implemented in several industry applications.

The third class of DFT algorithms also rely on path sensitization but instead of reconfiguration or retiming, uses circuit techniques to redesign the operational amplifier core [47] to permit more control inputs to sensitize propagation paths. This class is only recently introduced but the ATPG algorithm is similar to that discussed above. Many follow-on works on switched operational amplifiers [48] [49] [50] have shown significant improvements in design topologies and area reduction. It is expected that the area overhead for the switched operational amplifier structures would be lower than adding multiplexing switches. The performance issue still needs more study but shows promise that the input stages present lower loading than other comparable DFT structures.

The fourth class of DFT strategies reverts to functional test, using built-in self-test (BIST). On-chip circuits to generate proper test inputs and to measure some characteristics of the outputs are provided, and the pass/fail decision is made strictly based on the measured output. A representative work [51] discusses a BIST system for a sigma-delta analog-to-digital converter, taking advantage of existing circuits to reduce layout areas of the test generator and the output comparator. There is no ATPG algorithm as such since only a limited number of tests can be performed and all the tests are simplified conventional functional tests.

Numerous other BIST strategies have been proposed for analog circuits, with their own test strategies involving specific ATPG algorithms to exploit the built-in test nature of the techniques. [52] and [53] employ analog scan, using voltage buffers to mimic the digital scan flip-flop and thus digital scan-based test generation algorithms may be modified to apply to this structure. [54] and [55] also employ analog scan but use current mirrors to transmit signals. [56] uses on-chip circuits to test on-board resistors or capacitors for tolerance, and discusses strategies to test multiple digital-to-analog converters using resources on chip as well as those provided by a tester. [57] and [58] add a large number of components (comparator, subtractor, autozero circuit, counter) to implement converter functional tests on chip without having to use ATPG.

A recent DFT technique does away with ATPG altogether by incorporating extra circuit element to induce oscillation of the circuit under test in the test mode [59]. The frequency of oscillation of a circuit under test is related to component values or to other circuit parameters. The sensitivity of this frequency to the component under fault can be maximized to select an appropriate test frequency and the most suitable oscillator structure. Once the structure is designed as part of the circuit under test, there is no need for further ATPG.

DFT-based analog ATPGs share several characteristics: They are very well defined and customized to exploit the nature of the added DFT circuits; they involve performance impact, and thus they have not been widely used in industry. However, considering the history of digital ATPG and DFT, it is expected that DFT-based analog ATPG algorithms will become an important factor in advancing the state-of-the-art analog and mixed-signal testing.

4.4 Conclusions

It is interesting to note that the four classes of analog ATPGs have many common features:

1. The fault model is mostly parametric, with a few using catastrophic faults as a supplement. The model derivation varies slightly: a few use defect statistics to

generate the component variations, many use tolerance windows. The model generation procedure thus is mixed between functional and structural.

2. Most algorithms actually approach exponential complexity. This is quite an interesting feature, considering that analog circuits are small in terms of device count and number of inputs and outputs. The complexity issue emerges in the optimization or search algorithm to find a test, and while several heuristics have been proposed, none has found wide acceptance.

3. Most ATPG techniques have not been experimentally validated over a sufficiently large class of circuits to provide convincing evidence of their validity in general applications.

The future of analog ATPG, or more properly speaking, mixed-signal ATPG, depends on how these issues are resolved. There must be, for each circuit class, a commonly accepted fault model that covers at least structural faults (including parametric faults due to structural defects) and validated with respect to conventional specifications. The algorithm complexity is reduced by various techniques: divide-and-conquer with ATPG for smaller macros, path-sensitization methodologies to permit propagation of test signals between macros, and DFT structures (BIST or bus-based or well-defined structures) to control and observe internal signals. In digital systems, a subset of analog systems, DFT is required to reduce complexity, and it is only reasonable to expect that some form of DFT will be required as well to mitigate the analog ATPG complexity. Experimental validation is, in the final analysis, critical to the user's acceptance of any analog ATPG. After all, test, especially analog test, is an experimental discipline and unless the measurements validate the algorithms, the published works are academic with relatively little impact on the future of analog test.

Acknowledgment   This chapter is a reprint of an article that appeared in the IEE Proceedings G: Circuits and Devices in December 1996, and has been reprinted with their permission.

References

1. M. Soma, "Fault coverage of DC parametric tests for embedded analog amplifiers," presented at IEEE Int. Test Conf., Baltimore, MD, 1993.

2. "Semiconductor Industry Technology Workshop conclusions," presented at Semiconductor Industry Technology Workshop, 1993.

3. A. J. Bishop and A. Ivanov, "On the testability of CMOS feedback amplifiers," presented at Int. Workshop on Defect and Fault Tolerance in VLSI Systems, 1994.

4. C. L. Wey and R. Saeks, "On the implementation of an analog ATPG: The linear case," *IEEE Trans. Instrum. Meas.*, vol. IM-34, pp. 442-449, 1985.

5. D. M. W. Leenaerts and J. van Spaandonk, "A contribution to testing of analog integrated circuits in the DC domain," presented at European Test Conference, Rotterdam, The Netherlands, 1993.

6.   C.-L. Wey, S. Krishnan, and S. Sahli, "Test generation and concurrent error detection in current-mode A/D converters," *IEEE Trans. CAD Integrated Circuits Syst.*, vol. 14, pp. 1291-1298, 1995.

7.   T. Zwemstra and G. P. H. Seuren, "Analog Test Signal Generation on a Digital Tester," presented at European Test Conference, Rotterdam, The Netherlands, 1993.

8.   T. Wei, M. W. T. Wong, and Y. S. Lee, "Analog element-level fault diagnosis based on large change sensitivity computation in the frequency domain," presented at IEEE Int. Mixed Signal Testing Workshop, Quebec City, Canada, 1996.

9.   R. Ramadoss and M. L. Bushnell, "Test generation for mixed-signal devices using signal flow graphs," presented at IEEE Int. Conf. VLSI Design, Bangalore, India, 1996.

10.   A. Balivada, J. Chen, and J. A. Abraham, "Efficient testing of linear analog circuits," presented at IEEE Int. Mixed Signal Testing Workshop, Grenoble, France, 1995.

11.   A. Balivada, J. Chen, and J. A. Abraham, "Analog testing with time response parameters," in *IEEE Design & Test of Computers*, pp. 18-25, 1996.

12.   S.-J. Tsai, "Test vector generation for linear analog devices," presented at IEEE Int. Test Conf., Nashville, TN, 1991.

13.   S. Mir, M. Lubaszewski, V. Kolarik, and B. Courtois, "Optimal ATPG for analogue built-in self-test and fault diagnosis," presented at IEEE Int. Mixed Signal Testing Workshop, Grenoble, France, 1995.

14.   R.-W. Liu, "Selected papers on Analog fault diagnosis," in *Advances in Circuits and Systems*. New York: IEEE Press, pp. 143, 1987.

15.   N. B. Hamida and B. Kaminska, "Analog circuit testing based on sensitivity computation and new circuit modeling," presented at IEEE Int. Test Conf., Baltimore, MD, 1993.

16.   N. Nagi, A. Chatterjee, A. Balivada, and J. A. Abraham, "Fault-based automatic test generator for linear analog circuits," presented at IEEE Int. Conf. Computer-Aided Design, Santa Clara, CA, 1993.

17.   F. Corsi, M. Chiarantoni, R. Lorusso, and C. Marzocca, "A fault signature approach to analog devices testing," presented at European Test Conference, Rotterdam, The Netherlands, 1993.

18.   H. H. Schreiber, "Fault dictionary based upon stimulus design," *IEEE Trans. Circuits Syst.*, vol. CAS-26, 1979.

19.   A. Abderrahman, E. Cerny, and B. Kaminska, "Effective test generation for analog circuits," presented at IEEE Int. Mixed Signal Testing Workshop, Grenoble, France, 1995.

20.   S. Slamani and B. Kaminska, "Multifrequency testability analysis for analog circuits," presented at IEEE VLSI Test Symp., Princeton, NJ, 1994.

21.   N. B. Hamida and B. Kaminska, "Multiple fault analog circuit testing by sensitivity analysis," *Analog Int. Circuits & Sig. Proc.*, vol. 4, pp. 231–243, 1993.

22.   E. Felt and A. L. Sangiovanni-Vincentelli, "Testing of analog systems using behavioral models and optimal experimental design techniques," presented at IEEE Int. Conf. Computer-Aided Design, San Jose, CA, 1994.

23.   K. Saab, D. Marche, B. Kaminska, N. B. Hamida, and G. Quesnel, "LIMSoft: Automated tool for sensitivity analysis used for test vector generation," presented at IEEE Int. Mixed Signal Testing Workshop, Grenoble, France, 1995.

24.   N. B. Hamida, K. Saab, D. Marche, B. Kaminska, and G. Quesnel, "LIMSoft: Automated tool for design and test integration," presented at IEEE Int. Mixed Signal Testing Workshop, Quebec City, Canada, 1996.

25.   M. J. Marlett and J. A. Abraham, "DC_ITAP — An iterative analog circuit test generation program for generating DC single pattern tests," presented at IEEE Int. Test Conf., Washington D.C., 1988.

26.   G. Devarayanadurg and M. Soma, "Analytical fault modeling and static test generation for analog ICs," presented at IEEE Int. Conf. Computer-Aided Design, San Jose, CA, 1994.

27. G. Devarayanadurg and M. Soma, "Dynamic test signal design for analog ICs," presented at IEEE Int. Conf. Computer-Aided Design, San Jose, CA, 1995.

28. A. Richardson, A. Bratt, I. Baturone, and J.-L. Huertas, "The application of I_{DDX} test strategies in analogue and mixed signal ICs," presented at IEEE Int. Mixed Signal Testing Workshop, Grenoble, France, 1995.

29. G. Gielen, Z. Wang, and W. Sansen, "Fault detection and input stimulus determination for the testing of analog integrated circuits based on power-supply current

30. J. S. Beasley, H. Ramamurthy, J. Ramirez-Angulo, and M. DeYong, "I_{DD} pulse response testing of analog and digital CMOS circuits," presented at IEEE Int. Test Conf., Washington, DC, 1993.

31. Y. Miura, "Real-time current testing for A/D converters," in *IEEE Design & Test of Computers*, pp. 34-41, 1996.

32. S. S. Somayajula, E. Sanchez-Sinencio, and J. Pineda de Gyvez, "A power supply ramping and current measurement based technique for analog fault diagnosis," presented at IEEE VLSI Test Symp., Atlantic City, NJ, 1994.

33. C.-Y. Pan and K.-T. Cheng, "Pseudo-random testing and signature analysis for mixed-signal circuits," presented at IEEE Int. Conf. VLSI Design, San Jose, CA, 1995.

34. L. Milor and V. Visvanathan, "Detection of catastrophic faults in analog integrated circuits," *IEEE Trans. CAD Integrated Circuits Syst.*, vol. 8, pp. 114-130, 1989.

35. L. Milor and A. L. Sangiovanni-Vincentelli, "Optimal test set design for analog circuits," presented at IEEE Int. Conf. Computer-Aided Design, Santa Clara, CA, 1990.

36. L. Milor and A. L. Sangiovanni-Vincentelli, "Minimizing production test time to detect faults in analog circuits," *IEEE Trans. CAD Integrated Circuits Syst.*, vol. 13, pp. 796-813, 1994.

37. T. M. Souders and G. N. Stenbakken, "Cutting the high cost of testing," in *Spectrum*, pp. 48-51, 1991.

38. S. D. Huss and R. S. Gyurcsik, "Optimal ordering of analog integrated circuit tests to minimize test time," presented at IEEE Design Automation Conf., 1991.

39. W. M. Lindermeir, H. E. Graeb, and K. J. Antreich, "Design-based analog testing by characteristic observation inference," presented at IEEE Int. Conf. Computer-Aided Design, San Jose, CA, 1995.

40. V. Kaal and H. Kerkhoff, "On the optimization and optimal selection for analog/mixed-signal macros," presented at IEEE Int. Mixed Signal Testing Workshop, Quebec City, Canada, 1996.

41. J. L. Huertas, A. Rueda, and D. Vazquez, "Design for testability techniques applicable to analog circuits," presented at European Test Conference, Rotterdam, Holland, 1993.

42. J. L. Huertas, A. Rueda, and D. Vazquez, "Improving the testability of switched-capacitor filters," *Analog Int. Circuits & Sig. Proc.*, vol. 4, pp. 199–213, 1993.

43. J. L. Huertas, A. Rueda, and D. Vazquez, "Testable switched-capacitor filters," *IEEE J. Solid State Circuits*, vol. 28, pp. 719-724, 1993.

44. D. Vazquez, A. Rueda, and J. L. Huertas, "A new strategy for testing analog filters," presented at IEEE VLSI Test Symp., Atlantic City, NJ, 1994.

45. M. Soma, "A design-for-test methodology for active analog filters," presented at IEEE Int. Test Conf., Washington DC, 1990.

46. M. Soma and V. Kolarik, "A design-for-test technique for switched-capacitor filters," presented at IEEE VLSI Test Symp., Cherry Hill, NJ, 1994.

47. A. H. Bratt, R. J. Harvey, A. P. Dorey, and A. M. D. Richardson, "Design-for-test structure to facilitate test vector application with low performance loss in non-test mode," *IEEE Elect. Letters*, vol. 29, pp. 1438-1440, 1993.

48. D. Vazquez, A. Rueda, J.-L. Huertas, and A. M. D. Richardson, "Practical DFT strategy for fault diagnosis in active analog filters," *IEEE Elect. Letters*, vol. 31, pp. 1221-1222, 1995.

49. D. Vazquez, J.-L. Huertas, and A. Rueda, "Reducing the impact of DFT on the performance of ana-

log integrated circuits: Improved sw-opamp design," presented at IEEE VLSI Test Symp., Princeton, NJ, 1996.

50. D. Vazquez, A. Rueda, and J.-L. Huertas, "Fully differential sw-opamp for testing analog embedded modules," presented at IEEE Int. Mixed Signal Testing Workshop, Quebec City, Canada, 1996.

51. M. F. Toner and G. W. Roberts, "A BIST scheme for a SNR, gain tracking, and frequency response test of a sigma-delta ADC," *IEEE Trans. Circuits & Syst. II: Analog & Digital Sig. Proc.*, vol. 42, pp. 1-15, 1995.

52. C. Wey, "Built-in self-test (BIST) structure for analog circuit fault diagnosis," *IEEE Trans. Instrum. Meas.*, vol. 39, pp. 517–521, 1990.

53. L. T. Wurtz, "Built-in self-test structure for mixed-mode circuits," *IEEE Trans. Instrum. Meas.*, vol. 42, pp. 25–29, 1993.

54. M. Soma, "Structure and concepts for current-based analog scan," presented at IEEE Custom Integrated Circuits Conf., San Jose, CA, 1995.

55. C.-L. Wey and S. Krishnan, "Built-in self-test (BIST) structures for analog circuit fault diagnosis with current test data," *IEEE Trans. Instrum. Meas.*, vol. 41, pp. 535-539, 1992.

56. M. S. Nejad, L. L. Sebaa, A. Ladick, and H. Kuo, "Analog built-in self-test," presented at IEEE VLSI Test Symposium, Philadelphia, PA, 1994.

57. J. Rzeszut and B. Kaminska, "A built-in self-test method for digital-to-analog converter," presented at New Frontiers in Test Workshop, Montreal, Canada, 1993.

58. K. Arabi, B. Kaminska, and J. Rzeszut, "A new built-in self-test approach for digital-to-analog and analog-to-digital converters," presented at IEEE Int. Conf. Computer-Aided Design, San Jose, CA, 1994.

59. K. Arabi and B. Kaminska, "Oscillation-test strategy for analog and mixed-signal integrated circuits," presented at IEEE VLSI Test Symp., Princeton, NJ, 1996.

Design for Test

Bapiraju Vinnakota and Ramesh Harjani

In many instances, traditional approaches to test generation and application may be too expensive. Alternatively, test generation may not provide an adequate coverage of expected physical defects. Design for test is one approach to reduce total test costs and/or improve defect coverage[1-3]. A design for test (DFT) technique modifies a circuit to improve its testability. A DFT scheme can improve testability in a variety of ways such as by improving fault coverage, reducing test generation costs, or reducing test application times. In this chapter, we will focus on DFT for fault coverage improvement and test generation time reduction. These modifications usually affect operation only during circuit test. In general, the modifications require the addition of on-chip hardware and may impact circuit performance. The term test synthesis is used to refer to DFT schemes incorporated into an automated synthesis process. Note that unlike a BIST scheme, a DFT technique does not eliminate the need to generate tests and observe the results of tests. However, the overhead is usually less than with a BIST scheme. The benefit of a DFT scheme is judged by the improvement in a target metric, such as the stuck-at fault coverage (for digital circuits). The cost of a DFT scheme is the percentage increase in circuit area, and any performance impact. Though their primary aim is to reduce test costs, DFT schemes lead to costs and benefits at several different phases of the design and manufacturing process. For example: test synthesis schemes have been used to assist in design debug and verification; the increase in die size leads to a decrease in process yield. Thus, though they are accepted as being useful, it is challenging to precisely quantify the benefits of a DFT scheme.

The rest of this chapter is organized as follows. In the next section, we discuss essential preliminaries. First, we discuss how the differences between analog and digital circuits impact the formulation of DFT schemes. In Section 5.2, we discuss generic DFT techniques that are not restricted in their application to a specific class of circuits. Recent research efforts have concentrated on developing macro-specific DFT techniques. Macro-specific techniques reduce overhead costs by exploiting the characteristics of specific types of circuits. Section 5.3 contains a detailed review of DFT schemes for common macros such as filters and data converters. The last section focuses on techniques to evaluate the ability of DFT schemes to detect parametric faults, especially multiple parametric faults.

5.1 Preliminaries

The main source of test difficulties in digital circuits are circuit size and complexity[1]. Digital DFT techniques are designed to improve the ability to the control and observe the values of internal signals. The most common and well-known digital DFT technique is scan design. In a scan scheme, all memory elements are linked to form a shift register in the test mode. Thus, all internal memory elements are directly controllable and observable at the primary inputs and outputs. Other schemes improve testability in different ways. Some examples will be useful. The divide-and-conquer approach partitions a circuit into smaller sub-circuits each of which is tested independently. Additional circuitry and control are required to achieve this partitioning during circuit test. A second approach, used in large systems such as microprocessors, use special-purpose test buses. The test buses are used to access and observe internal subsystems. A third approach is to control and observe a specific set of internal nodes. One commercial DFT scheme makes almost all internal nodes observable by transporting their signal values to the primary outputs through a dedicated metal layer. As with other aspects of testing, many researchers have attempted to extend digital DFT techniques to analog and mixed-signal circuits. However, the unique characteristics of analog and mixed-signal circuits, discussed in detail in Chapter 1 and in other chapters as well, make it difficult to directly extend digital DFT techniques.

5.1.1   Analog Characteristics

While a comprehensive review of the unique characteristics of analog circuits is redundant, a brief discussion of the characteristics of analog circuits that impact the development of DFT schemes will be beneficial.

- The lack of an adequate fault model is a significant impediment. To compensate, many researchers have developed fault models specifically for DFT scheme validation. The drawback of this approach is that comparative evaluation of DFT schemes is very difficult. Secondly, evaluations of the abilities of many schemes to detect parametric faults have been limited.

- Digital DFT schemes typically have to only target spot defects. Spot defects can be modeled adequately using the single fault model. Thus, assessing a DFT scheme by the gains in single stuck fault coverage is meaningful. Analog test has to target both spot defects and parametric faults. The latter can only be modeled with a multiple fault model. Since the number of multiple faults is large, multiple fault coverage can usually only be quantified statistically.

- Many digital DFT techniques depend on an ability to transport internal signals to the primary outputs. Analog signals cannot easily be transported. A simple technique will lead to errors, while a reliable technique may be expensive. In other words, internal signals are difficult to measure accurately.

- Design techniques in analog circuits are usually specific to a class of circuits. Similarly, most DFT techniques developed by researchers are circuit specific, or class- or macro-specific. In a large circuit, composed of several macros, test techniques for individual macros are combined for the circuit as a whole.

5.1.2   Common Characteristics

DFT methods for analog and mixed-signal circuits have adapted to these constraints. Most DFT schemes for analog circuits aim to reduce test generation difficulties and improve fault coverages. They require the addition of hardware for switching circuitry and functional elements. Current macro-specific DFT techniques fall into one of the following categories:

Reconfiguration for test   A system's testability can be improved by incorporating the ability to arrange its components in more than one configuration during test. Reconfiguration techniques require the addition of a switching network, and the creation of a special test mode in which the circuit is reconfigurable. Reconfiguration has been used to improve testability in analog and mixed-signal circuits in two ways. We will refer to the first technique as *reconfiguration for access*. This technique is based on the divide-and-conquer approach that has been successfully employed in large digital systems. The circuit to be tested is partitioned into several blocks such that the inputs to and the outputs from each component can be directly controlled and observed. When it is targeted for test, a block is effectively isolated from the rest of the system. The formation of blocks is guided by tools which analyze the testability of the circuit and identify those circuit nodes which may be difficult to control and/or observe. The details of the application of reconfiguration for access are circuit specific. However, the structure of a target component sub-circuit is not altered for test. The subcircuit tested is identical to the one used during normal operation. Reconfiguration simply increases access to the circuit components, but does not alter them.

The second reconfiguration-based DFT technique is an indirect approach to testing. This approach to DFT is unique to analog circuits. Reconfiguration is used to alter the structure of a circuit to make it easier to test. We will refer to this technique as *reconfiguration for redesign*. Reconfiguration is used to rearrange the components to form an easily testable circuit. Even after reconfiguration, individual components may still not be directly accessible from the primary inputs and outputs. The response of the altered circuit, in its easily testable configuration, is used as a guide to determine if the original circuit, in its normal configuration, will work. For example, in one approach reconfiguration is used to verify that the values of all circuit components are at their expected values. The computed component values are then used to determine that the circuit, in the original configuration of components, will meet specifications. Unlike circuit-specific reconfiguration for access techniques, the reconfiguration for redesign techniques discussed in the literature have been macro-specific.

Code-based test   This approach is based on methods used to design on-line error detection schemes for fault-tolerant digital systems. When fault tolerance is required errors in system data have to be detected and corrected without interrupting operation. Data errors cannot be detected on-line by continually measuring all signals in a system. Redundant data codes are used to resolve this measurement problem. This idea has been extended to analog and mixed-signal circuits by several researchers. Code-based techniques target the difficulty encountered in measuring the values of on-chip signals. A redundant data code is used to encode on-chip data. On-chip code checkers, rather than off-chip instrumentation, can then be used to verify that the data code is not corrupted. Faults are detected indirectly in that faults that corrupt the code are the only ones to be flagged by the code checkers. To obtain a high fault coverage, the circuit can be redesigned to improve the number of faults that corrupt the data code. To limit overhead, the codes used have been very simple.

5.2 Generic Test Techniques

In spite of the lack of general design techniques, and the significant differences between various types of circuits, several "generic" test methods have been developed [4-10]. We refer to them as generic techniques since they usually do not exploit any properties of specific classes of circuits. Nor are they restricted in their application to a specific class of circuits.

5.2.1   Increased Controllability/Observability

The general goal is to increase the ability to control and observe the values of internal signals [4]. Reconfiguration for access is one method to achieve this goal. In the context of analog and mixed-signal circuits, components are isolated for test at the macro level. The circuit is partitioned during test such that all macros, such as data converters, are directly controllable and observable. In digital circuits, multiplexers are used to increase the ability to control and observe internal signals. Similarly, here, internal macros can be isolated by using analog multiplexers. However, several buffers may have to be added to ensure that no signals are loaded as a result of the modifications. Observability can also be increased by adding several direct metal contacts which are wired to the signals of interest. Macro-based testing is made more difficult in the context of feedback, which is present in most analog circuits. To be effective, macro-based testing requires significant involvement from designers. One advantage is that this approach is compatible with current design practices for analog circuits.

A second approach to increasing controllability/observability at the circuit level is to provide the ability to control/observe specific internal nodes. To minimize overhead, the number of additional "test-point" signals controlled/observed should be minimized by intelligent selection. When specifications are used as tests, the number of tests to be applied can be quite high. The problem of minimizing test time by reducing the number of tests, and appropriately ordering them, is related to test-point selection. Several test-time reduction techniques such as integer linear programming [10] and probabilistic analysis [50] and matrix triangularization for models based on sensitivity analysis [51,52], have been investigated. The basic approach in most of these techniques is to form a linear matrix model for the device under test using sensitivity analysis. For each test point, the model relates variations in component parameter values to variations in the output values. A choice of techniques is then available to identify the set of "significant" tests. Some techniques select all test points simultaneously, while others select them iteratively. The sensitivity matrix can be altered to accommodate the additional controllability/observability offered by test-points. Test selection algorithms [10,50,51,52] can be used as test-point selection algorithms.

5.2.2   A/D Boundary Control

In mixed-signal systems, the DFT scheme used for digital components can be used to test the analog components. The digital signals at the analog/digital interface in a system can be made part of a scan chain, as shown in Figure 5-1. The scan chain can be used to apply stimuli and observe the responses of the analog component.

In [5] test time is reduced by 40% by including scan-based techniques for a DSP-based mixed-signal telecommunications chip. The chip included a medium resolution high-speed charge redistribution A/D, a medium resolution high-speed charge redistribution D/A and two high resolution low-speed D/A converters. The digital portion of this chip was verified using scan techniques including an A/D replacement mode to circumvent the A/D converter. The outputs from the high-speed A/D and inputs to the high-speed D/A were directly interfaced to scan chains. This allowed for complete parametric test of these two modules. The inputs to the low-speed D/As were directly accessed by an internal bus controlled by the DSP core. Frequency domain tests were run to evaluate the performance of these converters. A similar scan-based approach for a mixed-signal ASIC library was also proposed in [8]. The outputs of an 8-bit A/D module were tied to the scan chain. All the modules in this library were designed such that they could be individually isolated. Once isolated predefined application-independent test vectors can be used to test each module.

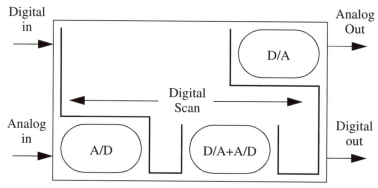

Figure 5-1 *Reconfiguration for direct access.*

Providing additional flexibility to the digital scan cells at the analog/digital boundary can make the analog component easier to test. For example, the dynamic capabilities of a data converter can only be tested by applying several inputs sequentially. Rapidly applying multiple inputs sequentially in a large digital scan chain is not a trivial task. A new scan-design called Ascan [6] proposes additional programmability for scan elements on the analog/digital boundary. The important benefit offered by this scan technique is that multiple data inputs/outputs can be applied/observed serially at the analog/digital boundary, more quickly than with a normal scan chain. The scan chain is designed such that only the analog components, and not the entire circuit, are exercised during this process. Lastly, in a long digital chain, the cells corresponding to data converter inputs/outputs can be collectively accessed as words. When applied to an analog processor Ascan resulted in the test time being halved.

5.2.3   System-Specific Test Techniques

In systems with a significant DSP core, the digital component can be used to generate tests for D/A converters and observe outputs from A/D converters. The existence of a DSP core reduces DFT specific overhead and permits complex signal analysis. In [9] the DSP core is used to generate the input stimulus for and to analyze the output response of the data converters. In this implementation a 13-bit D/A and a 13-bit A/D are tested using a 24-bit DSP core. All the digital circuitry is assumed to have been previously tested using pseudo random patterns, march patterns, and checker patterns. Because of the availability of the complete DSP core and the low resolution requirement a complete FFT analysis was feasible. Frequency domain tests were performed to extract the SNR and the effective number of bits. An external "gold" D/A converter is first used to evaluate the on-chip A/D converter. This is followed by a digital-to-digital measurement that evaluates the on-chip D/A. Because of the already existing DSP the necessary hardware overhead is less than 0.5% of the total chip area. Additionally, the total test time is reduced to 10% of the time necessary using traditional test equipment to perform parametric test.

5.2.4   Analog Scan

Given their success, scan techniques have also been developed for analog circuits. In analog circuits capacitors form the memory of a circuit. Scan techniques are limited to observing internal signal values. With analog scan techniques one cannot control the state of a circuit. Scan has been implemented using either bucket-brigade devices or charge coupled devices. In either implementation, the scan is executed in two phases. Initially, the values of internal signals are captured and held. The captured signals are then transmitted to the outputs using an analog shift register. In bucket-brigade implementations, sample and hold circuits are used to capture an internal signal. A chain of capacitors and voltage-follower buffers are used to form an analog shift register. In a CCD-based scheme, charge converters are used to capture internal voltages and currents and convert them to charges. A CCD register is used to shift the captured charges to the circuit outputs. In both implementations, care has to be taken to ensure that the test hardware does not impact the circuit being modified. Analog scan techniques have been covered in detail in [3].

5.2.5   Boundary Scan

At the board level, test buses provide access to board components and interconnect. This DFT technique requires multiplexers, and switches at the I/O boundary of the board components. These switches permit direct control and observation of a component's inputs and outputs and/or interconnect. Hence, this technique is usually referred to as boundary scan. In a device modified for boundary scan, a scan cell is added at each input/output pin. The 1149.1 (digital) boundary scan standard specifies the design of boundary cells and the protocol used in the test mode. As at the circuit level, boundary scan cells also form a shift register in the test mode. The standard is also compatible with self-test techniques within individual ICs. Many analog test bus techniques have been motivated by the digital standard. Most schemes suggested are compatible with the existing standard. A boundary scan cell is added to each input and output pin. In addition to permitting normal operation, each cell has to test the interconnect wire leading to or away from the pin as well as its own component. Recently, a draft of a proposed standard passed first ballot. The new standard is described in complete detail in a separate chapter.

5.3 Macro-Based DFT

Many complex functions, such as data conversion and filtering, are common to several analog systems. Though circuit-level realizations of these functions are not unique, all the realizations share a common set of specifications and have several structural commonalities. For several of these functions, specific realizations have been shown to be good. With minor modifications, these realizations, referred to as macros, are commonly used. Several researchers have developed DFT techniques for specific macros. Macro-specific DFT schemes exploit unique characteristics of either the functionality or the structure of the target macro. A system-level DFT scheme can be formed by combining macro-level DFT schemes. Efforts have concentrated on popular macros such as op amps, data converters, and filters.

5.3.1   Operational Amplifiers

Operational amplifiers (op amps) are the most common smaller analog modules found in mixed-signal ICs. Op amps are rarely used in the open-loop configuration, except as comparators. The performance of an in-circuit op amp is completely dependent on the feedback network around the amplifier. This implies that it is extremely difficult to develop a generalized DFT scheme for op amps. As with most other analog modules, op amps are characterized primarily by I/O performance parameters. Examples include low frequency gain, unity gain bandwidth, phase margin, load characteristics, input referred noise, offset voltage, input bias current, maximum slew rate, input common-mode range, output voltage range, power supply rejection ratio (PSRR), common-mode rejection ratio (CMRR), area and total power consumption [3]. Normal testing involves measuring a number of these parameters and is extremely time-consuming. Both code-based and reconfiguration-based DFT schemes have been developed for op amps.

5.3.1.1   Code-Based DFT

A fully-differential circuit is defined as one which has differential inputs, generates differential outputs and maintains a fully-differential signal path throughout the circuit. That is, all nodes that are on the signal path are members of a differential signal pair. In an analog IC, the absolute accuracy of circuit components is poor. However, the relative matching between components is extremely good. Hence, fully-differential analog circuits provide the highest performance in integrated environments. The penalty paid is an increase in circuit area. Differential amplifiers are used in high-performance applications including switched-capacitor filters, transconductance-C filters, and sigma-delta converters.

Consider a general differential circuit. The two input signals to a fully-differential circuit are also a differential signal pair. Let the two input voltages to a differential circuit be V_{in}^{1} and V_{in}^{2}. Then:

$$V_{in}^{1} = V_{in}^{bias} + V_{in}^{ss} \qquad (5.1)$$

$$V_{in}^{2} = V_{in}^{bias} - V_{in}^{ss} \qquad (5.2)$$

V_{in}^{bias} is the input bias voltage, generated by the preceding circuit, and V_{in}^{ss} is the input small-signal voltage. The two output voltages are also a differential signal pair. Usually, the output bias and the bias value for differential pairs internal to the circuit are derived from an internal or external voltage reference. As at the inputs and outputs, for every wire on the small signal path there exists another wire with the inverse small signal value. The input and output bias voltages are also referred to as the common mode components of the input and output respectively.

The data in differential circuits is inherently encoded [11]. We will refer to this code as the fully differential analog code (FDAC). Note that the bias voltage is also a component of the FDAC. This redundancy can be exploited for fault detection. A single fault inside the op amp will affect only one of the two signal propagation paths. The common mode component of the output FDAC will be altered by the fault. In other words, a single fault will corrupt the output FDAC. A code checker at the outputs of the op amp can be used to detect faults within the op amp. Fault coverage can be improved only by ensuring that a large number of faults corrupt the output differential signal. This will avoid the need to monitor multiple differential signals. The objective can be achieved by: (1) Replacing a single current source by two interacting current sources, (2) Choosing the common mode feedback structure intelligently. With these techniques almost all faults within an op amp will corrupt the output differential pair. The fault coverage is increased substantially. The structure of a code-based test scheme is shown in Figure 5-2.

Checker Design Figure 5-3 shows a design for an FDAC code checker. The checker monitors the common mode component of a differential signal. Each checker consists of two sub circuits. One sub circuit amplifies any detected shift in the common mode. The second latches this amplification to maintain a permanent record. To account for process variations and noise, the checker is not triggered until the common mode is at least some (preset) Δ_c away from the expected value. Here on, we will refer to the Δ_c as the checker tolerance. The impact of the checker threshold on the fault coverage will be discussed in Section 5.4.

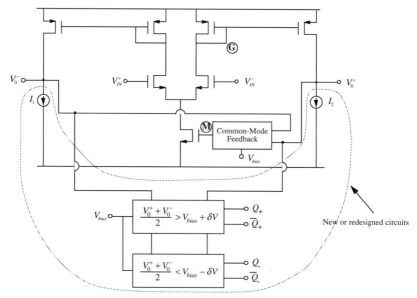

Figure 5-2   Differential op amp with checkers.(Figure courtesy of the IEEE, © 1994 [26].)

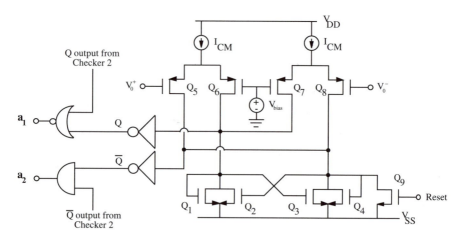

Figure 5-3   Output code checker for the differential op amp. (Figure courtesy of the IEEE, © 1994 [26].)

The area overhead of this technique is limited. The performance overhead is negligible (6% worst case). The checker consumes DC power but it can be turned off during normal operation. The analysis in this study focused on the impact of spot defects using inductive fault analysis. These spot defects led to both large (catastrophic faults) and small (parametric faults) component variations. All catastrophic faults and a large fraction of parametric faults (a total of 95% of faults in the op amp). However, the fault coverage for multiple minor component parametric faults was not evaluated. The extension of this code-based technique to the system level will be discussed in Section 5.3.3.1.

5.3.1.2   Reconfiguration-Based DFT

Both the techniques reviewed in this section are reconfiguration for redesign schemes. They alter the feedback structure around an in-circuit op amp during test.

Current Based Test   A reconfiguration technique for a supply current based test scheme is suggested in [13]. The technique is based on a bridging fault model specific to the study. A catastrophic bridging short fault model was developed for the op741 op amp. In a stand-alone configuration, it was shown that two simple DC tests could detect a large fraction of the modeled catastrophic faults. The remaining faults were detectable by monitoring other common output parameters such as the 3 dB bandwidth and the slew rate.

Unfortunately, in large circuits with embedded op amps, the op amps' performance is governed by feedback circuitry. Moreover, direct access to their inputs and outputs is limited. For an active filter, it was found that functional tests could not detect a large number of faults in embedded op amps. In the test mode, each op amp is configured as a source follower using additional switching circuitry as shown in Figure 5-4. A small resistor, either passive or active, is attached to the output of each op amp. In the source follower mode, as shown in Figure 5-4, the output voltage at each op amp is essentially the offset voltage. The current from the power supply is determined by the offset voltage and the resistor R. It was determined empirically that the offset voltage in a majority of the faulty op amps was substantially higher than the offset voltage in a good op amp. Thus, faulty op amps would draw more current from the power supply. The threshold for fault detection was set empirically. The fault model considered only catastrophic faults. The analysis of the impact of minor parametric variations was limited. Multiple parametric faults were not considered in the study.

Oscillation Based Test   Verifying the ability of a circuit to oscillate [12] has been suggested as another method to detect faults in op amps. In this approach, feedback circuitry is added to an op amp in the test mode such that the resulting circuit is an oscillator. The oscillation frequency is determined by circuit parameters such as the gain of the op amp, the unity gain frequency, and the values of the passive components. The core idea in the approach is that a fault in the op amp, or indeed in any component, will either prevent the circuit from oscillating or alter the oscillation frequency. A catastrophic fault model for 741 op amp was developed (for this study). A majority of the catastrophic faults resulted in loss of oscillation, because the poles of the circuit moved off the imaginary axis. Some faults preserved

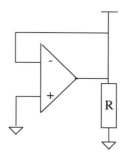

Figure 5-4 *Op amp configured as a source follower.*

the ability to oscillate, but the oscillation frequency changed substantially in the faulty circuit. A subsequent study investigated the impact of parametric faults. However, only single parametric faults were considered in the study. The application of this technique at the system level is discussed in Section 5.3.3.2.

5.3.2 Data Converters

Data converters transform analog signals to digital (A/D) and digital signals to analog (D/A). The conversion process approximates an analog signal by a finite resolution (n-bit) digital representation. A number of different topologies exist for data converters and these are usually classified by their rate of conversion. In general, the conversion rate and the resolution of the converter are inversely related. Most DFT schemes are designed for specific converter topologies. Hence, this review of DFT techniques for data converters is organized according to converter types.

5.3.2.1   *Traditional Test Techniques*

As with other analog and mixed-signal circuits, data converters have traditionally been tested using parametric techniques [14-21]. These tests verify functional performance parameters rather than the values of circuit components of the data converter. Data converters are characterized by their static and dynamic performance parameters. The static performance parameters characterize the input-output relationship of the converter at or near DC. The dynamic performance parameters characterize the converter at operating speed. All performance measures are concerned with the deviation of the transfer function from an ideal staircase [18], where the step height of the staircase is a function of the converter's resolution. Static measures have been tackled better by DFT techniques than dynamic parameters. Examples of static performance characteristics include the gain error, the offset error, the integral linearity and the differential linearity. Figure 5-5 shows some examples of the impact of errors in these parameters. Dynamic A/D converter performance characteristics include the signal-to-noise ratio (SNR), effective bits, aperture errors, and the input signal bandwidth. The SNR and effective bits are dynamic methods to measure the minimum resolution and identify errors in the transfer characteristics.

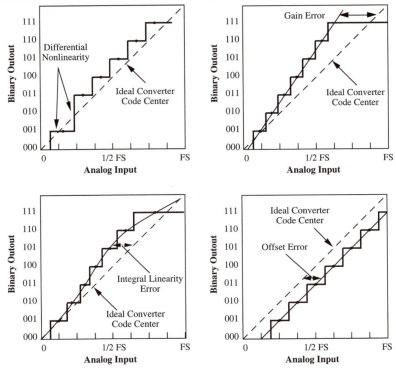

Figure 5-5 *Output parametric errors in A/D converters.*

Traditional test techniques consist of either directly measuring the parameters of interest or inferring them from statistical measurements. Direct methods include: comparing converter outputs to a "gold" standard, analog difference signal methods, crossplot methods, and servo loop code transition measurement methods. The code density test (or histogram test) measures the linearity of the converter by using a well-known and near full-scale input signal and evaluating the output code probability[18]. More recently, frequency domain techniques have been used to measure the harmonic content of the converted signal to provide an estimate of the SNR and effective number of bits [18,19,20]. In this technique a discrete-time Fourier transform is performed on the output data sequence and is used to measure converter performance characteristics. Next, we consider DFT techniques for specific data converter architectures.

5.3.2.2   Sigma-Delta Converters

Sigma-delta converters are oversampled converters that use digital filters combined with a low resolution D/A converter to generate extremely high resolution A/D converters. These converters are common components found in digital audio consumer products. The extremely high resolutions combined with the fact that they use both analog and digital circuits make sigma-delta converters hard to test. However, because of the in-built DSP core it is possible to develop fairly effective BIST schemes. In [22,23] a simple digital oscillator is combined with a sigma-delta D/A to generate an extremely high-quality analog sine wave. The sigma-delta loop moves the quantization noise to higher frequencies. A narrow-band digital bandpass filter is used to measure the signal output and a narrow-band notch filter is used to measure the noise level. This scheme tends to overestimate the SNR, however, it is easily compensated. The filter-based technique has the advantage of much lower overhead compared to a direct FFT-based technique to extract SNR results. The basic scheme has been extended to include frequency response and intermodulation distortion by incorporating multitone signal generators and multi-band filters[6]. This technique for BIST works particularly well with sigma-delta converters because of the large amount of on-chip DSP circuitry normally associated with such converters. These techniques are discussed in complete detail in the next chapter.

5.3.2.3   Successive-approximation Converters

Successive-approximation converters are medium speed converters and are one of the most common converter topologies. In CMOS, the charge redistribution implementation of the successive-approximation converter is most widely used. Both techniques to increase the observability of internal nodes and BIST techniques have been proposed for such converters.

Multiplexed A/D and D/A converters can often be used in medium speed telecommunications circuits. Such multiplexed data converters are usually implemented using a successive-approximation A/D architecture because such architectures use a D/A submodule which may be multiplexed to work as a separate D/A as well. In modern CMOS processes such successive approximation converters are designed using binary weighted capacitors based on charge redistribution techniques. The performance of such converters is limited by capacitor matching errors. In [24] a self-test scheme to test for such mismatch errors is described. The scheme uses an auxiliary high resolution sample-and-hold, an auxiliary coarse D/A converter to correct for offset errors, and a 1/2 LSB offset. The auxiliary sample-and-hold needs to be accurate only in terms of linearity as any offset errors are easily corrected by the auxiliary D/A. The circuit works as follows. A digital code is converted by the D/A. The internal D/A is operated in the inverting mode so that a single error in the A/D and D/A do not cancel each other. This signal is then held by a sample-and-hold until the data converter can be reconfigured as an A/D. Any offset for a zero input digital signal is corrected by the auxiliary D/A converter. The necessary input codes to test for a single mismatch error were also developed for this circuit.

A similar BIST technique for a D/A has also been suggested in [25]. The technique uses an additional sample-and-hold and an autozeroed window comparator. If the difference between the signal generated by the D/A subtracted from an external "gold" reference is greater than 1 LSB of the data converter, it is considered to be faulty. The technique can also be extended to successive approximation A/D converters as they use a D/A as a submodule within them. A BIST scheme for a switched-current successive approximation converter is discussed in detail in a separate chapter.

5.3.2.4   Pipelined Converters

Pipelined converters are extremely high-speed converters and offer performance comparable to flash converters but consume much less area and power. Since they offer several attractive features, they have received substantial attention from test researchers.

As the name implies, a pipelined converter is a multistage system. Each input is converted over several cycles. At any instant the converter concurrently converts several analog inputs, each in a different stage of the conversion process.

Architecture   In the basic architecture, each stage of a converter computes a single bit in the output word. An n-bit converter has n-stages. Let $2 \times V_{ref}$ be the maximum input voltage converted. A stage s_i producing bit b_i operates as follows:

1. Let the input voltage be V_i.
2. If $V_i > V_{ref}$, bit $b_i = 1$, else $b_i = 0$. This serves as single-bit ADC.
3. The output $V_{i+1} = 2(V_i - b_i V_{ref})$. That is, the analog value corresponding to b_i is subtracted and the result doubled.

In the general version of the architecture, a stage s_i converts k_i bits using a k_i bit A/D converter. A more complex D/A converter is required to subtract the analog voltage corresponding to the multibit digital output from the input. The resulting signal is multiplied by 2^{k_i} before being passed to the next stage. Thus, each stage consists of:

1. A k_i bit A/D converter.
2. A k_i bit D/A converter.
3. An amplifier with a gain of 2^{k_i}.
4. The output $V_{i+1} = 2^{k_i}(V_i - BV_{ref})$, where B is the k_i-bit output of the A/D converter in stage s_i.
5. A unit referred to as an MDAC (multiplying DAC) is used to accomplish operations 2 - 4.

Digital Error Correction   The resolution of a pipelined converter is limited by the mismatch errors in the capacitors used in each stage. Note that the residue, the difference between the input analog value and the digital bits produced in stage s_i cannot exceed V_{ref}. Any residue above this will exceed the conversion range of the next converter stage. If the converter operates as desired, the residue will not exceed V_{ref}. Because of capacitor mismatches, the digital code, produced by stage s_i, corresponding to an analog input may be incorrect. Consequently, the residue passed to the next stage will also be erroneous, and also exceed the conversion range of the next stage. A common solution is to incorporate redundancy into the conversion process [30]. The redundancy consists of three components:

1. The total number of data bits produced by all the stages is greater than the desired converter resolution. Thus $\sum k_i > N$, where N is the resolution.
2. The range of the stages in the converter is increased by decreasing the gain of the amplifier (usually halving it).
3. Using digital logic, the digital outputs of stage s_{i+1} are used to correct the outputs of stage s_i.

Unlike error correction techniques in digital systems, these error correction techniques cannot flag errors which are beyond their correction capability. Thus, digital error correction actually makes a converter harder to test, since it is difficult to set the inputs so as to test the correction logic [30]. In other words, digital error correction schemes are design techniques rather than DFT schemes. Thus, they are not discussed further in this chapter. Readers are referred to for more details on digital error correction and test techniques for such converters [30].

Virtual-Converter Test   The virtual converter DFT scheme is based on the structure of the pipelined converter[27]. It was developed for the single-bit-per-stage topology and uses the data duplication code to detect faults. Consider an n-stage converter. Let the stages be labeled $s_{n-1}s_{n-2}...s_2s_1s_0$. The n-1 most significan stages $s_{n-1}s_{n-2}...s_2s_1$ form an $(n$-$1)$-bit converter. Similarly, the n-1 least significant stages $s_{n-2}s_{n-3}...s_1s_0$ form an $(n$-$1)$-bit converter. In other words, one may form two virtual converters during the test process. If the same input is supplied to the two converters they should produce the same output. Notice that the two converters share n-2 physical stages. When both converters are working, each stage alternately "belongs" to each of the two virtual converters. The operation of the two virtual converters is shown in Figure 5-6. Though each stage computes a digital output on every clock cycle, the throughput of each of the virtual converters is half of that of the original converter.

It would appear that since the two converters share a substantial amount of hardware, faults cannot be detected. However, any shared physical stage calculates a different significant bit in each of the two virtual converters. A single physical fault in a shared stage causes errors in the outputs of both converters. However, the errors in the outputs of the two virtual converters are of different magnitude. A unique advantage of this technique is that all single faults in any stage, above a certain magnitude, can be proven to be detected by the virtual converter test technique. As two n-1-bit converters are being compared only defects that are larger than 2 LSB are detectable. However, with one extra stage an n-bit converter can be tested to n-bits of resolution. Any simple analog signal that spans the entire input range is sufficient to test for both parametric variations in device parameters and catastrophic faults.

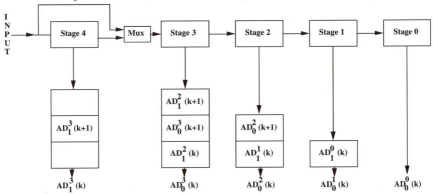

Figure 5-6 *Virtual converter test technique.(Figure courtesy of the IEEE, © 1997 [28].)*

Multiple minor parametric faults were not considered in the original study. However, subsequently, the coverage of multiple parametric faults that cause output errors was found to be about 60%. This technique offers the advantage of simplicity and very low overhead, a single switch, and perhaps a stage. In addition, each stage is operated at full frequency during test.

Spare Stage   A duplication code-based test based on the use of redundant hardware was suggested in [29]. This technique is aimed at both off-line test and concurrent error detection. It is applicable to converters which convert multiple bits per stage. A spare MDAC, a ROM containing test stimuli, a switching network, and two comparators are added to the converter. Each stage is tested using the following procedure:

1. First, the MDAC in each stage is tested using the spare MDAC. The gain of the MDAC is set to 1, that is, it is converted to a simple DAC. The ROM is used to supply test stimuli to the target stage and the spare. The analog outputs are compared using a comparator. The individual stages are tested serially.

2. The second test step relies on the idea that the output of a stage should not

exceed a threshold. For example, in a single-bit-per-stage converter, the output voltage should not exceed V_{ref}. A comparator is used to monitor the output during operation and verify that it does not exceed the reference threshold. In this phase, the MDAC is presumed to be working correctly.

This technique offers the advantage of simplicity. The validity of the technique was verified using a catastrophic fault set developed specifically for the study. The impact of single parametric faults was also simulated. However, the impact of multiple minor parametric faults was not considered. The overhead associated with this technique is considerable as the switching network will require substantial area.

5.3.3   Filters

Analog filters find widespread use in several applications. These filters may either be discrete time or continuous time filters. Switched capacitor filters are the most common discrete time filters. Traditional test techniques for filters are based on verifying their functional specifications [3,31]. These include verifying, among other parameters the: (1) Variation of the gain within the passband; (2) Location of the edge(s) of the passband; (3) Performance of the filter in the stopband; (4) Dynamic range, voltage range, output noise etc.; (5) Maximum input and clock frequencies. Verifying these functional specifications is a time-consuming process. Hence, substantial effort has been directed toward developing DFT techniques for IC filters.

Each of the efforts in this area exploits different characteristics of filter circuits. Several exploit the regularity in the structure of filters. Almost all filters use multiple stages, where each stage normally consists of an op amp and a feedback network composed of resistances (which may be switched capacitors) and capacitors. The regularity has been exploited to create multiple paths for data propagation and to increase access, but with low overhead costs. The DFT techniques reported in the literature are not specific to the functionality realized by the filter or the topology of the filter. They are usually applicable to both discrete and continuous time filters. Hence, unlike the review of DFT schemes for data converters, this section is organized according to the basic techniques employed. We start with a review of code-based techniques.

5.3.3.1   Code-Based DFT

Almost all code-based techniques are based on forming two different paths for data propagation in the test mode. This is usually achieved through the addition of hardware and/or switching circuitry. At the system level a code-based test scheme has to address the issue of identifying the signals to be monitored, as well as the number and location of code checkers.

In large circuits, monitoring only the circuit outputs with a code checker will not provide an adequate fault coverage. Recall that a fault is detected only if it corrupts the code in a set of signals that are monitored by a checker. In a multistage circuit, a fault in an internal stage may corrupt the code at several internal signals. However, this corruption may not be propagated to the system outputs. Thus, in a large circuit, several sets of signals have to be observed by a code checker. Either multiple checkers will have to be used, or all the coded signals of interest will have to be transported to a central location. Multiple checkers will increase the area overhead. Transporting signals and routing them through an analog multiplexer may cause an increase in the checker threshold. The solution chosen depends on the checker design. A checker that uses passive components is more expensive than a checker that uses active components. Thus, it is more expensive to replicate. However, in passive component checkers the parameters of the active components are far less important. Several designs for checkers are discussed in this chapter. A checker for a code-based DFT scheme will have to compare two signals V_1 and V_2 and determine if they are equal within some threshold Δ_c. Typically, two checkers are used, one to determine if $V_2 > V_1$, or vice versa (ignoring the threshold). However, only one checker that flags errors in one direction only need be used in most circuits. This is because errors produced by any test input can be inverted in sign by inverting the small signal test inputs. We start with a review of circuits which naturally contain two paths for data propagation.

Differential Circuits   The inherent redundancy in differential circuits can also be exploited at the system level[32-35]. As discussed above, the code at multiple sets of signals have to be monitored. One option is to place checkers at the outputs of each stage in the filter, that is at the outputs of the op amp in that stage. Recall that faults can be detected when they corrupt the fully differential analog code (FDAC). Faults internal to an op amp will corrupt the FDAC code at its outputs. Next, consider a fault in a passive component that is a part of the feedback circuitry. This fault will alter the common mode voltage at the inputs to at least one op amp (stage). A stage is designed to suppress any common mode voltage at its inputs. Therefore, FDAC errors at the inputs to a stage will not be propagated to its outputs. Thus, errors in the passive components can only be detected by placing checkers at the inputs to each stage in the filter. Input checkers at stage s will also detect errors in the active components of the stages whose outputs are fed to stage s. In summary, in a multistage filter the input common mode voltages to each stage have to be monitored. These may be monitored with multiple checkers, or by transporting the signals of interest to a central location.

Checker Design   Figure 5-7 shows a design for an input common mode checker that uses active components[32]. In a differential pair, the common source node V_s tracks the ICMV linearly, but is offset by the quiescent gate-to-source voltage drop across the input devices Q_1 and Q_2. To eliminate this offset, a third diode-connected nMOS transistor, Q_3, is added to the differential pair. A measurement of the ICMV is then available at the drain of transistor Q_3. Area and power overhead is reduced by decreasing the size of Q_3 and scaling down its bias current by the same amount. In our checker, the width of Q_3 is only a fifth as large as the differential pair transistors. The tolerance of the checker is determined by simulating the impact of process variations. A checker that uses passive components is discussed in the next subsection.

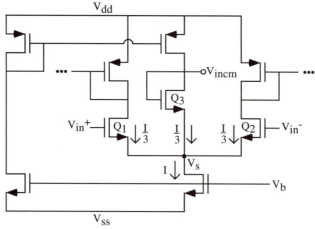

Figure 5-7 *Input common mode voltage checker. (Figure courtesy of the IEEE, © 1997 [28].)*

A fifth order low-pass filter using this DFT scheme was designed and fabricated. The design placed a checker at the inputs to each of the five stages. The total area overhead for the additional checkers was about 10%. The design included artificial fault injection to verify the quality of the DFT scheme. This technique detected all significant single faults that in both the active and passive components were injected artificially. Though the area overhead of this technique is very low, its ability to detect multiple parametric faults is limited. Methods to determine parametric fault coverage are discussed in the next section.

Spare Stage   This technique is identical in principle to the duplication code-based DFT scheme discussed for data converters. In a filter, the topologies of all the stages will be very similar. That is, stages may have the same basic structure but different component values. In [36,37] an additional programmable stage is added to the filter. In the programmable stage, the values of the capacitors can be altered (using switches). The programmable values are chosen such that the additional stage can replicate (in various settings) the functionality of each of the stages in the multistage filter. A block diagram of the test structure is shown in Figure 5-8. The test process is executed in several phases. In each phase:

1. The redundant stage is programmed to replicate the functionality of one of the stages of the filter.
2. The target stage's input is applied to the spare stage using the input selector.
3. The output of the target stage is selected with the compare selector and compared with the output of the spare.

A tolerance is incorporated into the comparator to account for process variations and phase shifts. A switching network is also added to allow the redundant stage to be programmed and compared with each stage in the filter. Any single fault is unlikely to affect both the target stage and the redundant stage. This technique has been implemented and fabricated for an IC filter. One drawback of the technique is its substantial overhead for the extra stage and the switching network, nearly 80%. However, the overhead is less than that required by full duplication, which would require an overhead greater than 100%. The additional switching network may have

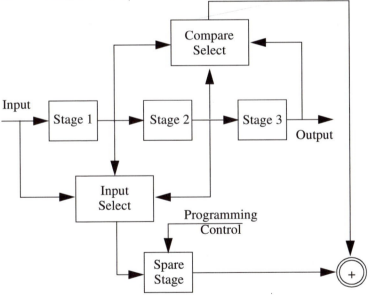

Figure 5-8 *Test using a programmable spare stage.*

a marginal impact on performance. The quality of this technique was verified only with respect to single catastrophic and parametric faults.

Pseudoduplication This technique targets single-ended switched capacitor filters which contain only a single path for data propagation[38]. In a typical filter, the passive capacitors occupy a far larger fraction of IC area than the active elements. Therefore, a spot defect is more likely to occur in a passive component, than in an active component. Further, the functionality of a filter is determined by the passive components. Pseudoduplication targets faults in the passive components and is based on two ideas: (1) The functionality of a switched-capacitor circuit is determined by capacitor ratios. Halving or doubling every capacitor on an IC will not alter the functionality of the filter; (2) Large capacitances are not realized as single blocks. Rather, for each design a "unit" capacitance is defined. Every capacitance is realized as a multiple of the unit capacitance, and is obtained by connecting the required number of unit capacitances in parallel.

Consider again the set of all capacitances on an IC. Assume that every capacitance can be partitioned into two equal halves during circuit test. Then, we would possess two sets of capacitances, both of which realized the functionality of the filter. The outputs produced by the "two" filters (one with each set of passive components) for the same inputs, should be equal. A single fault will affect only one of the two sets. The outputs of the two filters will not be equal in the presence of a fault. Two sets of capacitances can be realized, without actually duplicating hardware, in one of two ways:

1. Every unit capacitance can be realized as a pair of "half-units" that could be separated during the test process. In the test mode, the unit capacitance would be half its normal value. Parasitics would have a greater influence on the functionality. Figure 5-9 shows a method to fracture unit capacitances.

Figure 5-9 *Capacitance partitioning for pseudoduplication. (Figure courtesy of the IEEE, © 1997 [28].)*

2. The second approach is to preserve unit capacitances. Large capacitances which are even would be split into equal halves. In capacitances which are odd, only the "last" unit capacitance is split into two.

Though there are two sets of passive components, there is only one set of active components. One cannot realize two filters by alternating the active components between the two capacitor sets for the following reason. During the transition period there will be no feedback around the op amp and its output will saturate. The solution is to keep one of the two sets of capacitances permanently connected. The second set is switched in and out on alternate cycles. Thus, as with the virtual converter test technique, the two filters share hardware. Again, a single fault will affect the outputs of the two filters differently. The difference can be detected using a checker. When applied to a layout of fifth order low-pass filter, the two techniques resulted in overheads of 16% and 20% respectively. Pseudoduplication is effective at detecting catastrophic faults. However, with complete fracturing fault-free operation is also affected because of an increase in parastics.

Checker Design   Figure 5-10 shows a design for a checker that compares two voltages that are input on consecutive cycles. This design is a modified version of the ones presented in [35]. Let the input voltages at V_{in} on two consecutive cycles be V_1 and V_2. After cycle 2, the input to the comparator is at a voltage $V_2 - V_1$. If this is greater than 0, the comparator output is raised. In other words, this checker flags errors in one direction only, if $V_2 > V_1$. The overhead required for this design may be greater than that required for the previous design since a passive element is used. This design can be easily extended to the case when both the voltages to be measured are available simultaneously.

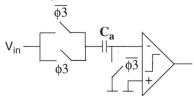

Figure 5-10   *Checker to compare two voltages over two successive cycles.*

Checksum Codes   Many analog circuits, especially filters, can be expressed as linear state-variable systems. The state of the circuit is stored in capacitors. Changes in the state and the value of the output are determined by the current state and current inputs to the circuit. In linear circuits, changes in the state can be computed using matrix-vector multiplications. In digital systems, checksum encoding is used to obtain on-line error detection and also correction in matrix-vector multiplication. This same technique is extended to analog circuits in [39]. Figure 5-11 shows an example of a checksum encoded matrix-vector multiplication. The last row in the matrix is the checksum vector. Each element in the checksum vector is the sum of the remaining elements in its column. Correspondingly, the bottom element in the output vector is the sum of the remaining elements in the output vector. Any single faulty operation will result in a corruption of the checksum code of the output vector. The costs of this code are the costs of the operations needed to compute the output checksum element. Multiple checksum vectors can be used to increase the error detection capability, and/or for error correction.

If the matrices used to compute the change in state are checksum encoded, the outputs of the multiplication will also be checksum encoded. The code checker verifies if the output actually obeys the checksum code. Again, since the signals are analog, unlike digital circuits, the comparison process has to incorporate tolerance. The overhead consists of that required to implement the checksum vectors, and the additional operations, in the circuit's functionality and to implement the checker. This technique was developed for on-line error detection. Though this technique can be used for off-line fault detection, its potential effectiveness is difficult to judge. As with differential circuits, only those faults which produce errors that corrupt the checksum code will be detected. It is not guaranteed that all faults will produce such errors, for an appropriate test input. Even if they do, no methods to generate an adequate test have been developed. If they do not, techniques to redesign circuits to improve fault coverage have to be developed.

$$\begin{bmatrix} 3 & 2 & 0 \\ 1 & 0 & 3 \\ 4 & 2 & 3 \end{bmatrix} \begin{bmatrix} 1 \\ 2 \\ 1 \end{bmatrix} = \begin{bmatrix} 7 \\ 4 \\ 11 \end{bmatrix}$$

Figure 5-11   *Example of a checksum code.*

5.3.3.2   *Reconfiguration-Based DFT*

Reconfiguration-based schemes rearrange the circuit components during test such that the resulting circuit is more easily testable. The circuit tested is not the one used during normal operation.

Programmable Biquads   Universal biquads are a two-stage filter structure that can be used to realize an arbitrary function of the form

$$\left(\sum_{i=0}^{2} a_i s^i \right) \bigg/ \left(\sum_{i=0}^{2} b_i s^i \right) \tag{5.3}$$

With one structure, universal biquads can be used to realize either a lowpass, highpass, or bandpass filter[40]. Figure 5-12 shows the structure of a universal biquad. In this DFT scheme, all multistage filters are realized using biquads. For filters with an odd number of stages, one of the component biquads would contain an unused stage. During normal operation, the biquads are programmed to execute the desired functionality. In the test mode the components are reconfigured and tested as follows:

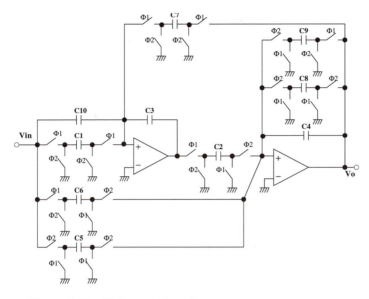

Figure 5-12 *Universal biquad.*

1. All the biquads are programmed to be constant gain stages. The biquad can be made a constant gain stage by setting each $a_{ij} = k_j b_{ij}$. The subscript j is used to refer to the coefficients in stage j. Then, the gain of stage j is k_j.

2. The reprogramming can be accomplished by altering capacitor values without altering the structure of the circuit. The constraints on the coefficients are used to identify the capacitors to be altered. As with pseudoduplication, capacitor values can be altered for test by switching in additional unit capacitors.

3. In the test mode, the gain of all the biquads in series is

$$\prod_{j=1}^{n} k_j = K \qquad (5.4)$$

4. An amplifier with gain $-K$ is added to the design. For the same input, the outputs of the reconfigured filter and amplifier are added. A nonzero result indicates the presence of a fault.

The biquads can also be tested in isolation to permit limited fault diagnosis. The analysis of this technique was limited to catastrophic faults. Parametric faults were not considered. The overhead required by the additional circuitry was of the order of 20 - 30%. The need to use biquads imposes constraints on the realization and perhaps an area penalty.

Stage Isolation   A different approach targeting switched capacitor ICs is suggested in [41,42]. A similar technique is developed in [43]. The basic idea is to increase controllability and observability of internal signals through additional control of the switches in the circuit. Consider the filter stage shown in Figure 5-13. The switches have been labeled with two labels. Those marked S_s transmit signals. Those marked S_g connect nodes to ground. If during test, those marked S_s are *ON* permanently, and those marked S_g are *OFF* permanently, then, the output voltage is simply an amplified (or attenuated) version of the input voltage. The ratio is determined by capacitor values, as shown in Figure 5-13. That is, the output follows the input voltage. Here too, the stages are tested sequentially.

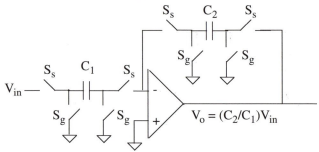

Figure 5-13   *Reconfiguration for stage isolation.*

When a stage is targeted for test, the switches in the remaining stages are set such that they function as voltage followers. The stages "before" the target stage are used to control the inputs to the target stage. Those stages "after" the target stage are used to observe the output of the target stage. Figure 5-14 shows an example configuration in the test mode. When the output of a stage is monitored with this technique, the interpretation of the observed values has to account for the path of signal to the output pin. The primary cost of this technique is the additional control for the switches, and additional control inputs to the IC. As with other techniques, this technique was verified only with respect to single catastrophic faults. The primary drawback is that there is no guarantee that faults which cause a filter to malfunction will be detectable when stages are tested in isolation. Conversely, faults which cause significant errors when a stage is operated in isolation may not cause significant errors at the system level.

Oscillation-Based Test   The oscillation-based technique, discussed for op amps in Section 5.3.1, can also be applied at the system level [44]. Given a filter, additional feedback is added such that the resulting circuit is an oscillator. Essentially, the additional feedback elements have to move the poles of the system onto the imaginary axis. The feedback is design-specific. It has been suggested that in

Figure 5-14 *Isolation-based test strategy.*

many cases, the additional circuitry required is minimal. As an example, as shown in Figure 5-15 only a single high-gain inverter and a switch are required to convert a low pass filter to an oscillator. A fault in either the active or passive components can have one of two effects:

1. The fault prevents the circuit from oscillating.
2. The oscillation frequency is shifted away from its good circuit value.

For two filters this technique was shown to detect a large number of catastrophic faults in the passive components. The effects of process variations on the oscillation frequency were simulated. The analysis of parametric faults was limited to single parametric faults. However, multiple parametric faults, which are more important than single parametric faults, were not addressed. The costs of this technique are the additional hardware and control logic.

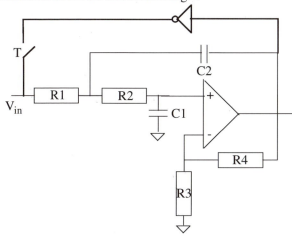

Figure 5-15 *DFT for oscillation-based test.*

Component Verification   This approach is based on the observation that capacitor ratios are more important than the absolute values of the capacitors in a switched capacitor circuit [45]. If all relevant capacitor ratios on a chip can be verified, the transfer function of the filter can be computed, and one may determine if the filter meets design specifications. As with the isolation test strategy, stages are reconfigured such that the input/output relationship is determined by the ratio of only two capacitors. Consider the circuit shown in Figure 5-16. Let P and I be non-overlapping clocks. When P = 1, the output $V_o = (C_2/C_1)V_{dd}$. If $C_1 > C_2$, then $V_o < V_{dd}$. By measuring the output, one may verify if the ratio C_2/C_1 is at its expected value. The capacitor C_1 is referred to as the integrating capacitor. We will refer to C_2 as the input capacitor. Based on this relationship, a test procedure, structured as follows, was proposed:

1. The output of every op amp is made directly observable in the test mode.
2. Every capacitor in the circuit is paired with at least one other capacitor.
3. The capacitor C_1 should be the larger one in each pair.
4. A feedback capacitor may be used in more than one pair.
5. In each test mode one ratio is verified at each op amp in parallel.
6. As shown in Figure 5-16, the supply voltage is input to each op amp.
7. Observing the output verifies the capacitance ratio.

The number of modes is determined by the maximum number of ratios to be verified at any op amp. To be testable, every capacitor has to be switchable in the test mode. Even capacitors which are not switched during normal operation have to be switchable in the test mode. Switching is necessary since the input voltage in the test mode is a constant. In addition, the input capacitor in each pair has to be connected to V_{dd} in the test mode. This eliminates the need for a separate test input voltage. The tolerances with which the ratios are measured determines the tolerance with which the specifications are estimated. In all cases, the outputs can be measured with a fixed tolerance. The corresponding tolerance in the capacitor ratio is determined by the value of the capacitor ratio. For example, if the precision in voltage measurement is $1mV$, for a supply voltage of $5V$, a ratio of 5 can be measured with a tolerance of 0.1%. A ratio of 50 can only be measured with a tolerance of 1%. Since the capacitor ratio is not a constant over all pairs in the test mode, a limitation is that not all ratios are measured with the same tolerance.

Additional switches and control are needed to pair capacitances in the test mode and to connect the input capacitance to V_{dd}. An analog multiplexer is used to make all op-amp outputs observable. The significant benefit is that all component values are observed directly. The cost of this technique is the overhead needed for reconfiguration and observability. An overhead of 10% was reported for a biquadratic filter. A more significant problem is that in many cases extra switches are inserted in the signal path, though they can be designed to minimize performance impact. Lastly, since the filter is not tested in its normal configuration, the impact of parasitics on normal operation cannot be judged.

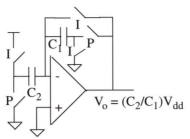

$V_o = (C_2/C_1)V_{dd}$

Figure 5-16   *Reconfiguration for component verification*

5.4 Quality Analysis

A test process is judged by its cost and the quality of the product that passes the test process. Almost all DFT techniques discussed in Section 5.3 have been shown to be effective at detecting locally catastrophic faults. That is, faults which cause significant variations in the value of a single component. The validation has been both through circuit simulation and through the fabrication of test ICs, some with artificial fault injection. In analog circuits, parametric faults, that is minor component variations, are as important as catastrophic faults. Some researchers have considered "parametric" faults in validating their DFT schemes. However, these efforts have been limited in several ways. In one study, parametric faults were defined as faults which caused minor variations in the output. In another, validation was limited to verifying the impact of minor deviations in the value of a single component. Both models of parametric faults are not sufficiently comprehensive. A precise definition of parametric faults will be beneficial.

Parametric Faults   Parametric faults cannot be defined with respect to the output. Parametric faults occur because normal process variations cause the values of one or more components to deviate from their expected value. Recall that some tolerance in output values is allowed. Minor single component deviations will usually not cause the output to be erroneous. That is, the deviation of the output from its expected value may be within acceptable tolerances. A collection of minor variations may cumulatively cause the output to be erroneous. However, not every possible set of variations is guaranteed to cause an output error. Only those multiple deviations that cause an output error constitute parametric faults. Simulation is usually the only method by which one may determine if a specific set of multiple component variations is indeed a parametric fault. The number of potential parametric faults is infinite. The fault coverage for multiple parametric faults cannot be judged by fabricating a small number of ICs, or by simulating a small number of faults. Probabilistic techniques are needed to analyze the impact of multiple parametric faults. Probabilistic techniques have previously been used to analyze general test application in analog and mixed-signal circuits[49,50]. We will adapt and extend these methods to analyze design for test techniques.

5.4.1   Preliminaries

In this subsection, we will define the metrics used to quantify the performance of a test process, and the models to be used for analysis. Some, but not all, are based on the concepts defined in [49,50].

5.4.1.1   Component/Output Models

Process statistics can be used to estimate the expected range of variation in the value of each on-chip component. Random variables are commonly used to represent and analyze the impact of process variations. We use Gaussian distributions[50] to model the impact of process variations on component values.

Components   In analog ICs, though absolute values can vary significantly, relative matching between components is high [46,47]. Thus, most analog ICs use the relative matching between similar components to set circuit parameters. Several examples of such circuits have been discussed in this chapter. When ratios are used to set transfer functions, output parameters are immune to shifts in the means of component values. Only relative shifts are of importance. Correspondingly, we usually need only model relative shifts in component values. Thus, on-chip components can be modeled by Gaussian variables centered around a zero mean. The σ of the distribution is determined by process statistics.

Outputs   In general, the output is a nonlinear function of the component values. It would appear to be difficult to compute the output distribution from the component distributions. However, at a test point, the transfer function, with respect to component values, can be linearized using sensitivity analysis[51,52]. The linear sum of several Gaussian distributions is also a Gaussian distribution[48]. Thus, about each test point the variations in the output value can also be modeled as a Gaussian distribution about a zero mean. For example, for the voltage follower in Figure 5-13, the variations in the output caused by minor variations in capacitor variations are described by the following equation.

$$\Delta V_o = \frac{\Delta C_2}{C_1} - \frac{C_2 \Delta C_1}{C_1^2} \tag{5.5}$$

Signal Tolerances   At each test point, a range of output values is defined as being acceptable. The output tolerance Δ_0 is the accepted variation in the output at a test point. Thus, the value of signal x_1 is acceptable if $-\Delta_0 < x_1 < \Delta_0$. The output tolerance will be circuit and test-point specific. Similar tolerances will have to be allowed in test points in a circuit reconfigured for redesign and in code checkers used to monitor on-chip signals.

5.4.1.2   Performance Metrics

A good IC is one for which all outputs are at their expected values. Let P_G be the probability of manufacturing a good IC. $P_B = 1 - P_G$ is the probability of manufacturing a faulty IC. An ideal test process will:

- pass all good ICs.
- fail all bad ICs.

In a real test process, some good units will fail the test, and some bad units will pass the test. Some, but not all, of the following terms, or terms similar to those listed below, have been previously defined [50]. The effort in [50] methods to reduce test time in analog circuits. We reuse these definitions in the context of design for test.

Definition   The following terms categorize good and faulty units with respect to their performance on the test process:

1. P_{GP} is the probability that a good IC will pass the test.
2. P_{GF} is the probability that a good IC will pass the test.
3. P_{BP} is the probability that a good IC will pass the test.
4. P_{BF} is the probability that a good IC will pass the test.

Definition   The fault coverage of the test process is $F = P_{BF}/P_B$.

Definition   The success rate is $S = P_{GP} + P_{BF}$.

Definition   The yield is $Y = P_{GP}/P_G$.

Definition   The confidence-level is $C = P_{GP}/(P_{GP} + P_{BP})$.

5.4.2   Analysis

To assess their quality, we use simple models for code-based and reconfiguration-based DFT schemes. The models are not meant to represent all code-based and reconfiguration-based DFT schemes in complete detail. However, they do represent essential characteristics of simple, but representative, reconfiguration-based and code-based DFT schemes.

5.4.2.1   Code-Based DFT

Our model for code-based DFT is based on the data duplication code, the most common code. Assume that the duplication code is employed at the circuit outputs. This model can be extended to other codes without significant difficulty. Let s_1 and

s_2 be the two signal outputs that are being compared. Refer to the entire s_1s_2-plane, represented in Figure 5-17, as U.

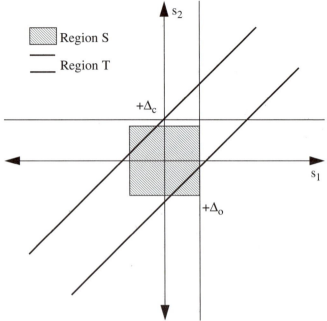

Figure 5-17   *Model for duplication code-based DFT.*

1. The circuit is acceptable only if both the outputs are at their expected values (within tolerance).
2. The on-chip checker is triggered if the two signals being compared are not equal (within tolerance).

Let the acceptable tolerances in the two outputs be Δ_o. Since the signals are identical, the expected values for the two outputs are identical as are the tolerances. Let the tolerance in the code checker be Δ_c. The output tolerance is specified by the user or design specifications and the code tolerance is specified by the test process. These ideas may be expressed in equations as follows.

The original circuit is good if:

$$-\Delta_o \le s_1, s_2 \le \Delta_o \tag{5.6}$$

That is, the original circuit is good only if both its outputs are near their expected values. This relationship is represented by the striped region in Figure 5-17. Any parametric deviations which cause both the outputs to be in this region will not be parametric faults. Refer to this area as region S. The circuit will pass the test, that is the code checkers will not be triggered, if:

$$-\Delta_c \le s_1 - s_2 \le \Delta_c \tag{5.7}$$

This relationship is represented by the area between the solid dark lines in Figure 5-17. Refer to this area as region T.

5.4.2.2   Reconfiguration-Based DFT

A similar modeling technique can be used for reconfiguration-based DFT. Consider a circuit with two components c_1 and c_2 operated in one configuration and reconfigured for test. Refer to the entire c_1c_2-plane, represented in Figure 5-18, as U. Consider two test points:

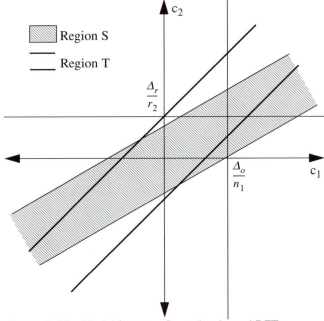

Figure 5-18   *Model for reconfiguration-based DFT.*

1. A specification test point O_{TP} in the original circuit that determines whether the circuit is good or bad.
2. A test point R_{TP} in the reconfigured circuit that determines whether the circuit passes or fails the test process.

At each test point, the output is a linear function of the same set of components. However, the equations for the original and reconfigured circuits will be different. Let the acceptable tolerances in the original and reconfigured circuits be Δ_o and Δ_r. The former is specified by the user or design specifications, the latter by the test process. These ideas may be expressed in equations as follows.

The original circuit is good if:

$$-\Delta_o \leq n_1 c_1 + n_2 c_2 \leq \Delta_o \qquad (5.8)$$

This relationship is represented by the striped region in Figure 5-18. Refer to this area as region S. Any set of parameter deviations that correspond to points in the striped region are not parametric faults. The reconfigured circuit will pass the test if:

$$-\Delta_r \leq r_1 c_1 + r_2 c_2 \leq \Delta_r \qquad (5.9)$$

This relationship is represented by the area between the solid dark lines in Figure 5-18. Refer to this area as region T.

5.4.3   Analysis

The information in Figures 5-17 and 5-18 demonstrates that with both types of DFT schemes, a test process may not function as desired. In the two figures, one may observe that:

- A portion of the unstriped region lies outside the solid lines. This is the region $(U - A) \cap (U - T)$, and it corresponds to the set of bad circuits which fail the test.
- A portion of the striped region lies within the solid lines. This is the region $A \cap T$, and it corresponds to the set of good circuits which pass the test.
- A portion of the striped region lies outside the solid lines. This is the region $A \cap (U - T)$, and it corresponds to the set of good circuits which fail the test.
- A portion of the unstriped region lies between the solid lines. This is the region $(U - A) \cap T$, and it corresponds to the set of bad circuits which pass the test.

The first two outcomes are the only desirable ones. The last two are undesirable outcomes of the test process. The quality of the test process is determined by the likelihood of each of these events occurring. The probability of each of these events occurring is determined by the probability density functions of the distribution of component and signal values. Theoretical analysis is possible for simple structures. Large complex circuits require empirical analysis.

5.4.3.1   Theoretical Techniques

With either reconfiguration technique, the probability of a specific event occurring can be computed by integrating the probability density functions for the components or signals over the appropriate region in the U plane. For example, in the circuit with two components discussed above, the probability of a bad circuit passing the test is (since component values are independently distributed):

$$P_{BP} = \int_{((U-A) \cap T)} p(c_1)p(c_2)dc_1dc_2 \qquad (5.10)$$

Similar expressions can be derived for code-based techniques. The probability of a good circuit failing the test in a duplication-based code is:

$$P_{GF} = \int_{(A \cap (U-T))} p(s_1,s_2)ds_1ds_2 \qquad (5.11)$$

If the signals in the duplication code are produced by disjoint hardware, then $p(s_1,s_2) = p(s_1)p(s_2)$. Similar expressions can be derived for other events such as P_{GP} and P_{BF}, for both types of DFT techniques.

5.4.3.2   Statistical Techniques

Clearly, the modeling and analysis methods described above are not restricted to circuits with two components, or to circuits employing the duplication code. Expressions to compute event probabilities of interest for circuits with n_c components and/or n_s signals can be derived by simple extension. It is usually not possible to numerically compute the corresponding integrals of joint density functions over multiple components for large complex circuits. Empirical analysis is necessary. Figure 5-19 shows the typical flow of a tool for Monte-Carlo analysis.

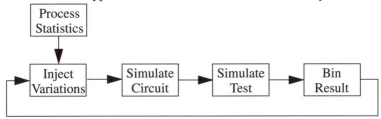

Figure 5-19   *Statistical parametric fault analysis.*

The target circuit is simulated N times. Process statistics are used to determine the range of variations in component values. On each iteration, the circuit is simulated with a different set of component values. The simulation determines if the circuit is good/bad and if it passes/fails the test process. The result of the simulation is used to bin the simulation and increment one of four counters (one for each possible outcome). The probability of an event is the value of a counter for a bin divided by the total number of runs. For example, the probability of a bad circuit failing the test is:

$$P_{BF} = \frac{N_{BF}}{N} \qquad (5.12)$$

Other parameters can be calculated similarly. We have developed a parametric fault-effect analysis tool (pFEAT) for switched-capacitor circuits. pFEAT uses circuits input in the SWITCAP format, and uses it as the circuit simulator. The tool can be used to evaluate both code-based and reconfiguration-based DFT schemes.

5.4.3.3   Results

In this section, as an example, we quantify the benefits of both code-based and reconfiguration-based DFT techniques when applied to the standard integrator shown in Figure 5-20.

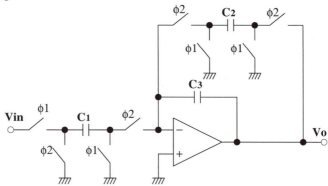

Figure 5-20   *A switched-capacitor integrator.*

The code used is the duplication code. We assume two disjoint copies of the integrator produce the signals being compared. The DC gain is the parameter tested. The event probabilities of interest, discussed in the previous section, were calculated using numerical integration. They are plotted in Figure 5-21. The *x-axis* is the ratio $m = \Delta_c/\Delta_o$. Performance parameters derived from these probabilities are plotted in Figure 5-22.

For reconfiguration-based test, the integrator is reconfigured as a voltage follower, by controlling the switches appropriately. The specification test point for the original circuit is the 3 dB frequency of the integrator. The same test point is used for the reconfigured circuit. Event probabilities were estimated using Monte Carlo simulation. A 1,000 trials were conducted. The results are shown in Figure 5-23. The *x-axis* is the ratio $m = \Delta_r/\Delta_o$. Performance parameters derived from these probabilities are plotted in Figure 5-24.

The information in the figures indicates that not all parameters of the test process can be optimized simultaneously.

• For low values of *m,* an unacceptably large fraction of good units fail. The output confidence level and fault coverage are high. However, the yield is low since too many good units fail the test.

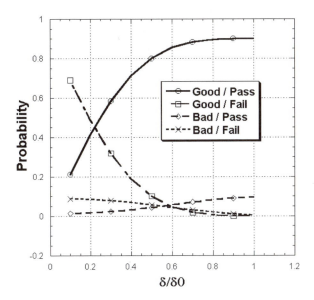

Figure 5-21   *Event probabilities for duplication code-based DFT.*

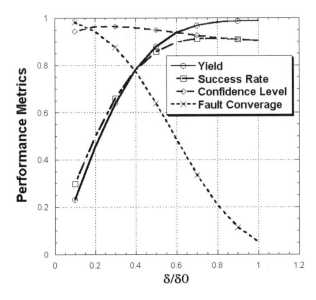

Figure 5-22   *Performance metrics for duplication code-based DFT.*

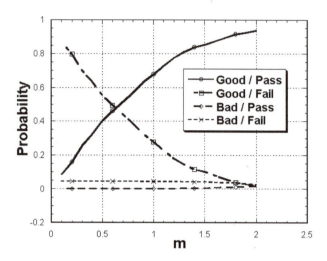

Figure 5-23 *Event probabilities for reconfiguration-based DFT.*

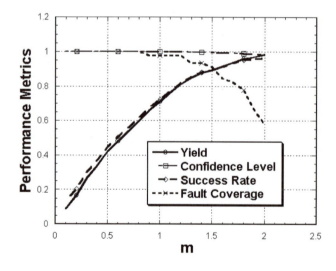

Figure 5-24 *Performance parameters for reconfiguration-based DFT.*

- For large values of m, an unacceptably large fraction of bad units pass the test. The fault coverage is low and the output confidence level is adversely affected. The yield is high since all good units pass the test.
- The total success rate is optimal for values of m between the two extremes. The optimal value will change if the definition of the success rate is changed,

that is, if we assume different relative costs for test process success and failure.

In summary, with either code-based or reconfiguration-based DFT it will be difficult to provide a high fault coverage for multiple parametric faults at a reasonable cost. The examples discussed in this section are not unique in this aspect. Monte-Carlo simulation has shown that similar results are produced with other types of codes and reconfiguration schemes. This may be explained as follows. Consider a general circuit with m specifications. In a general circuit, the region of acceptable component or signal values A is defined by the specifications. Again, let U refer to the space of all component or signal values. Let A_i be the region of the plane that passes specification i. Then, the region that corresponds to acceptable units is:

$$A = \bigcap_{i=1}^{m} A_i \qquad (5.13)$$

In a test scheme with k tests let T_i be the region in U plane that passes test i. Then, the region that corresponds to units that pass the test process is:

$$T = \bigcap_{i=1}^{k} T_i \qquad (5.14)$$

If the tests are the specifications, then $A \equiv T$. However, in general, both code-based and reconfiguration-based DFT schemes do not test specifications directly. Hence, usually, $A \neq T$, and consequently $P_{GF} \neq 0$ or $P_{BP} \neq 0$, even if $T \subseteq A$ or vice versa. In other words, the test process is likely to have undesirable outcomes. The probability of such events, corresponding to undesirable outcomes, occurring is determined by the probability density functions of component variations. As discussed in Chapter 2, better process control will reduce the probability of multiple parametric variations resulting in faults.

5.5 Conclusion

A wide range of DFT schemes has been developed for analog and mixed-signal circuits. Some general techniques, such as increased controllability and observability have been direct extensions of digital techniques. Several others have targeted the characteristics that make analog circuits difficult to test. Such schemes are specific to macros such as filters. Many macro-specific DFT techniques are based on either the use of codes or circuit reconfiguration. Code-based techniques aim to reduce the difficulty of measuring on-chip signals. Reconfiguration-based techniques typically rearrange components so as to form an easily testable circuit. Most DFT techniques proposed in the literature are very effective at detecting catastrophic faults or hard defects. Multiple parametric faults are of greater significance in analog circuits than catastrophic defects. Probabilistic analysis indicates that, because code or reconfiguration-based DFT techniques do not directly verify specifications, many DFT techniques may not guarantee a high fault coverage of multiple parametric faults at a reasonable cost.

References

1. M. Abramovici, M. Breuer, and A. Friedman, *Digital Systems Testing and Testable Design*, Englewood Cliffs, NJ, Computer Science Press, 1990.

2. R.-wen Liu, *Testing and Diagnosis of Analog Circuits and Systems,* Van Nostrand Reinhold, 1991.

3. M. Ismail and T. Fiez, eds., *Analog VLSI: Signal and Information Processing,* McGraw-Hill, 1994.

4. P. Fasang, D. Mullins, and T. Wong, "Design for testability of mixed-analog/digital ASICs," in *IEEE Custom Integrated Circuits Conference*, pp. 16.5.1-16.5.4, 1988.

5. P. Astrachan, T. Brooks, J. Everett, W. Law, K. McIntyre, C. Nguyen, and C. Weng, "Testing a DSP-based mixed-signal telecommunications chip," in *IEEE International Test Conference*, pp. 669-677, 1992.

6. J. Verfaillie and D. Haspeslagh, "A general purpose design-for-test methodology at the analog-digital boundary of mixed-signal VLSI," *Journal of Electronic Testing: Theory and Application*, vol. 9, pp. 109-115, 1996.

7. L. T. Wurtz, "Built-in self-test structure for mixed-mode circuits," *IEEE Transactions on Instrumentation and Measurement*, vol. 42, pp. 25-29, February 1993.

8. E. Flaherty, A. Allan, and J. Morris, "Design for testability of a modular, mixed-signal family of VLSI devices," in *IEEE International Test Conference*, pp. 797-804, 1993.

9. E. Teraoka, T. Kengaku, I. Yasui, K. Ishikawa, T. Matsuo, H. Wakada, N. Sakashita, Y. Shimazu, and T. Tokuda, "A built-in self-test for ADC and DAC in a single-chip speech CODEC," in *IEEE International Test Conference*, pp. 791-796, 1993.

10. F. Novak, I. Mozetic, M. Santo-Zarnik, and A. Biasizzo, "Enhancing design for test of active analog filters by using CLP(r)," *Analog Integrated Circuits and Signal Processing*, vol. 4, pp. 215-229, 1993.

11. B. Vinnakota and R. Harjani, "The design of analog self-checking circuits," *IEEE International Conference on VLSI Design*, 1994.

12. K. Arabi and B. Kaminska, "Oscillation-test strategy for analog and mixed-signal integrated circuits," in *14th IEEE VLSI Test Symposium*, pp. 476-482, April 1996.

13. C.-Y. Pang, K.-T. Cheng, and S. Gupta, "A comprehensive fault macromodel for op amps," in *IEEE International Conference on Computer Aided Design*, 1994.

14. J. R. Naylor, "Testing of digital/analog and analog/digital converters," *IEEE Transactions on Circuits and Systems*, vol. CAS-25, pp. 526-538, July 1978.

15. J. Doernberg, H.-S. Lee, and D. A. Hodges, "Full-speed testing of A/D converters," *IEEE Journal of Solid-State Circuits*, vol. SC-19, pp. 820-827, December 1984.

16. M. Mahoney, *Tutorial DSP-Based Testing of Analog and Mixed-Signal Circuits*, IEEE Computer Society Press, 1987.

17. B. E. Boser, K. P. Karman, H. Martin, and B. A. Wooley, "Simulating and testing oversampled analog-to-digital converters," *IEEE Transactions on Computer-Aided Design*, vol. 7, pp. 668-674, June 1988.

18. M. J. Demler, *High-Speed Analog-to-Digital Conversion*, Academic Press, 1991.

19. K. W. Hejn and R. C. Morling, "A semifixed frequency method for evaluating the effective resolution of A/D converters," *IEEE Transactions on Instrumentation and Measurement*, vol. 41, pp. 212-217, April 1992.

20. L. Benetazzo, C. Narduzzi, C. Offelli, and D. Petri, "A/D converter performance analysis by a frequency-domain approach," *IEEE Transactions on Instrumentation and Measurement*, vol. 41, pp. 834-839, December 1992.

21. A. Charoenrook and M. Soma, "Fault diagnosis of flash ADC using DNL test," in *IEEE International Test Conference*, pp. 680-689, 1993.

22. M. F. Toner and G. W. Roberts, "A BIST scheme for SNR test of a sigma-delta ADC," in *IEEE International Test Conference*, pp. 805-814, 1993.

23. M. F. Toner and G. W. Roberts, "A BIST technique for frequency response and intermodulation distortion test of a sigma-delta ADC," in *IEEE VLSI Test Symposium*, pp. 60-65, 1994.

24. C. A. Leme and J. Franca, "Efficient calibration of binary-weighted networks using a mixed analogue-digital RAM," in *IEEE International Symposium on Circuits and Systems*, pp. 1545-1548, 1991.

25. K. Arabi, B. Kaminska, and J. Rzeszut, "A new built-in self-test approach for digital-to-analog and analog-to-digital converters," in *IEEE International Conference on Computer Aided Design*, 1994.

26. R. Harjani and B. Vinnakota, "Analog circuit observer blocks," *Proc. VLSI Test Symposium*, pp., 1994.

27. R. Harjani and B. Vinnakota, "Analog circuit observer blocks," in *IEEE Transactions on Crcuits and Systems II*, pp. 258-263, 1997.

28. R. Harjani and B. Vinnakota, "ACOBs: A DFT Technique for Analog Circuits," Tutorial *Int Symp. Circuits and Systems*, 1997.

29. E. Perelias, A. Rueda, and J. L. Huertas, "A DFT technique for analog-to-digital converters with digital correction," in *15th IEEE VLSI Test Symposium*, pp. 302-307, Apr. 1997.

30. S.H. Lewis, R. Ramachandran and W.M. Snelgrove, "Indirect testing of digital-correction circuits in analog-to-digital converters with redundancy," *IEEE Trans. Circuits and Systems II: Analog and Digital Signal Processing*, vol. 42, no. 7, pp. 437-445, July 1995.

31. R. Gregorian and G. Temes, *Analog MOS Integrated Circuits for Signal Processing*, Wiley Series on Filters: Design, Manufacturing and Application, New York: Wiley and Sons, 1986.

32. N. J. Stessman, B. Vinnakota, and R. Harjani, "System-level design for test in analog circuits," *IEEE Journal of Solid-State Circuits*, pp. 1320-1326, October 1996.

33. V. Kolarik, M. Lubaszewski, and B. Courtois, "Towards self-checking mixed-signal integrated circuits," in *European Solid-State Circuits Conference*, 1993.

34. V. Kolarik, M. Lubaszewski, and B. Courtois, "Designing self-excersising analogue checkers," in *IEEE VLSI Test Symposium*, pp. 252-257, 1994.

35. S. Mir, V. Kolarik, M. Lubaszewski, C. Nielsen, and B. Courtois, "Built-in self-test and fault diagnosis of fully differential analogue circuits," in *IEEE International Conference on Computer Aided Design*, 1994.

36. J. L. Huertas, A. Rueda, and D. Vazquez, "Improving the testability of switched capacitor filters," *Analog Integrated Circuits and Signal Processing*, vol. 4, Kluwer Academic Publishers, pp. 199 - 213, 1993.

37. J. L. Huertas, D. Rueda, and D. Vazquez, "Testable switched-capacitor filters," *IEEE Journal of Solid-State Circuits*, vol. 28, pp. 719-724, July 1993.

38. R. Harjani, B. Vinnakota and W.-Y. Choi, "Pseudoduplication: an ACOB technique for single-ended circuits," in *Int. Conf VLSI Design*, Hyderabad, India, January 1997.

39. A. Chatterjee, "Concurrent error detection in linear analog and switched-capacitor state variable systems using continuous checkers," in *IEEE International Test Conference*, pp. 582-591, 1991.

40. D. Vazquez, A. Rueda, and J. L. Huertas, "A new strategy for testing analog filters," in *IEEE VLSI Test Symposium*, pp. 36-41, 1994.

41. M. Soma, "A design-for-test methodology for active analog filters," in *Proc. IEEE International*

Test Conference, pp. 183-192, 1990.

42. M. Soma and V. Kolarik, "A design-for-test technique for switched-capacitor filters," in *12th IEEE VLSI Test Symposium*, pp. 42-47, April 1994.

43. M. Renovell, F. Azais and Y. Bertrand, "The multi-configuration: a DFT technique for analog circuits," in *14th IEEE VLSI Test Symposium*, pp. 54-59, April 1996.

44. K. Arabi and B. Kaminska, "Parametric and catastrophic fault coverage of analog circuits in oscillation-test methodology," in *15th IEEE VLSI Test Symposium*, pp. 166-171, April 1997.

45. C. Dufaza and H. Ihs, "Test synthesis for DC test and maximal diagnosis of switched capacitor circuits," in *15th IEEE VLSI Test Symposium*, pp. 252-259, 1997.

46. J. B. Shyu, G. C. Temes, and F. Krummenarcher, "Random errors in MOS capacitors," *IEEE Journal of Solid State Circuits*, pp. 1070-1075, 1982.

47. M. J. M. Pelgrom, A. C. J. Duinmaijer, and A. P. G. Welvers, "Matching properties of MOS transistors," *IEEE Journal of Solid-State Circuits*, October 1989.

48. C. W. Helstrom, *Programability and Stochastic Processes for Engineers*, Macmillan Publishing Company, 1991.

49. J. Brockman and S. Director, "Predictive subset testing: optimizing IC parametric performance testing for quality, cost, and yield," *IEEE Transactions on Semiconductor Manufacturing*, pp. 104 - 113, 1989.

50. L. Milor and A. Sangiovanni-Vincentelli, "Optimal test set design for analog circuits," in *IEEE International Conference on Computer Aided Design*, pp. 294-297, November 1990.

51. J. Van Spaandonk and T. A. M. Kevenaar, "Selecting measurements to test the functional behavior of analog circuits," *Journal of Electronic Testing: Theory and Applications*, vol. 9, pp. 9-18, 1996.

52. G. N. Stenbakken and T. M. Souders, "Linear error modeling of analog and mixed-signal devices," in *IEEE International Test Conference*, pp. 573-581, 1991.

Spectrum-Based Built-in Self-Test

Benoit Veillette and Gordon Roberts

Built-in self-test has been accepted as a tool for digital verification and production test. It reduces the tester complexity, eliminates the need for off-chip interfacing, and allows the device to be tested many times during the manufacturing cycle of the product. In addition, through company standards, automated solutions can be created and integrated into present-day CAD facilities. A similar approach is now being attempted for the analog portion of a mixed-signal device. However, the tests that are performed on an analog circuit are quite different from those performed on a digital circuit. Digital circuits are normally tested for correct structure, for example, stuck-at faults, whereas analog circuits are tested to see if they perform the desired function within a range of acceptability. To perform such tests, accurate analog signal sources and measurement instruments are required. It is the intent of this chapter to describe a set of circuits that can be used to form an on-chip spectral-based measurement setup that is immune to process variations, requires no external trimming, and is silicon-area efficient. This scheme enables such measurements as signal-to-noise, gain-tracking, frequency response, and distortion to be performed directly on-chip.

6.1 Introduction

Most of the ideas for mixed-signal (MS) and analog verification are adaptations of digital testing methods. This is not surprising as digital test methods seem to be able to cope with the ever increasing logic density. While it is useless to blindly map digital solutions to the analog domain, the hindsight gained in digital testing should not be completely discarded for the MS and analog problem. For example, a concept that offers much is built-in self-test (BIST) [1]. While this acronym has been used for widely different schemes by researchers and CAD software companies, it generally implies a shift of the verification burden from the automated test equipment (ATE) to the IC. This is beneficial in a number of ways, the most evident being a simplified and cheaper ATE which leads to reduced test costs. Furthermore, signals do not need to travel long distances, reducing the load on the circuits. Another advantage, easily overlooked, is the possibility to implement a cradle-to-grave test strategy where the IC and its surroundings can be verified at every step of the manufacturing cycle in addition to the customer's premises. There is however a cost associated with BIST in the form of extra silicon area required for the stimulus generation, signature analysis, and control circuitry.

The specifications of analog circuits are much more complex than those of digital circuits. They thus require elaborate signal sources and parameter extraction instruments and long test times. Furthermore, the number of specifications is usually quite large. For digital verification, fast simulations and a simple fault model allow the test engineer to eliminate redundancy and thus test only a very small subset of the specifications. On the other hand, the need to take into account subtle process variations and the heavy computational demands of analog simulations make this weeding out process very difficult for any but the smallest analog circuits [2]. Attempts to simplify process variations into a simple fault model to reduce complexity may lead to incorrect solutions [3]. Any successful BIST scheme must thus provide measures that can be related to actual circuits specifications.

We propose a BIST scheme for the measurement of sinusoidal type or *spectral-based* metrics. These include frequency response (FR), gain tracking (GT), signal-to-noise ratio (SNR), and harmonic (HD) and intermodulation (IMD) distortion. The implementations we suggest are based on hardware-efficient analog signal generators which are entirely digital except for a 1-bit digital-to-analog converter (DAC). They therefore exhibit many digital circuit qualities such as programmability, repeatability, low sensitivity to process variations, and mature testing methods. The response of the circuit is analyzed in the digital domain. The presence of a digital output is thus essential. Analog devices that satisfy this criteria include voice-band CODECs, modems, charge-pump phase-locked loops, and digital wireless transceivers. Other circuits may also be verified provided they can be placed in series with a digital-output circuit. Several examples of this will be given later on in this chapter.

Section 6.2 will introduce the reader to some of the early BIST schemes for mixed-signal circuits and outline their short-comings. Subsequently, we shall introduce an alternative scheme, which, we believe, alleviates some of these limitations. Section 6.3 will describe the design of various analog signal generators available to designers for on-chip signal generation. We will describe in Section 6.4 several methods for on-chip parameter extraction. Section 6.5 will then outline various BIST schemes for baseband and narrowband systems. Finally, conclusions will be drawn and future work will be outlined.

6.2 Some Early BIST Schemes

We begin our survey with a technique which, at the date of this writing, seems to be the most appropriate for industrial use. This technique is a Built-In Self-Test for an ADC which was recently patented by AT&T Bell Laboratories [4]. A ramp voltage is generated and applied to the ADC input, and then the ADC output is checked for monotonicity and linearity using relatively simple digital circuitry. The ramp can be created by a circuit such as a charge pump commonly found in phase-locked loop circuits. However, the patent does not include a way to independently ensure the correctness of the ramp. This technique is very useful for measuring the Integral Non-Linearity (INL) and Differential Non-Linearity (DNL) of the ADC, but is less suitable for measuring dynamic performance such as frequency response and signal-to-noise ratio (SNR). Even so, this approach highlights several of the attributes which any BIST technique should have before industry would be willing to consider it seriously. First of all, it must be relatively straightforward to implement with simple on-chip circuits. Second, the output test result must be of a type that is widely used and recognized by manufacturers. Third, there must be no chance of fault "masking." Fault masking occurs when a faulty circuit escapes the test because its behavior was masked by a fault in the test circuitry or compensated by other devices tested concurrently. Finally, the addition of the circuitry to perform the BIST must not jeopardize the correct performance of the circuit. All proposals for BIST of a mixed-signal IC should be evaluated according to these criteria.

There have been many other suggested approaches. The classical loop-back test used by generations of transmission engineers has inspired several schemes for mixed-signal BIST. The HBIST technique proposed by Ohletz [5] is an interesting variation of this. His approach is intended to be used with mixed-signal ICs similar to the one depicted in Figure 6-1.

The idea is illustrated in Figure 6-2. A Pseudo-Random Bit Sequence (PRBS) generated by a Linear-Feedback Shift Register (LFSR) is fed into the digital side of the DAC. The analog output of the DAC is looped around into the analog input of the ADC. A second LFSR compacts the digital output of the ADC to generate a signature which is compared to a value stored in memory. This method is very similar to the signature compaction schemes used in digital BIST [6,7].

We can identify two concerns about this approach. The first is that it is possible for one fault in the DAC to "mask" or make unobservable a second fault in the ADC. For example, if the tester only injects a digital stimulus into the DAC and observes the digital ADC output, then an unacceptably low gain in the DAC can be

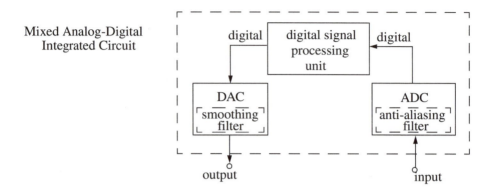

Figure 6-1   *A typical mixed-signal integrated circuit.*

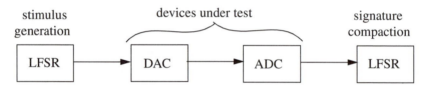

Figure 6-2   *Hybrid built-in self-test (H-BIST).*

masked by an unacceptably high gain in the ADC. The second issue is that using a deterministic comparison at the output leaves no allowance for the "fuzziness" or randomness that is present in every analog circuit. Analog testing normally involves a range of acceptable values; for instance, one may specify an acceptable gain of 20 dB with a 0.05 dB tolerance. The use of a signature comparison for the test implies that if thermal noise causes only the least significant bit of the ADC to change just once, the compacted signature will not match and the test will fail even though the circuit might still be acceptable. A very high rejection rate is the result. However, this signature-comparison technique is beautiful in its simplicity and may be useful for certain applications.

Recent BIST suggestions have taken into account the "fuzzy" nature of analog circuits. In [8], a simplified curve-fitting algorithm is presented that can be used to obtain a measure of an ADC non-linearity on-chip. Similarly, in [9], the authors present an analysis method based on histograms. The end-results are parameters which can be related to specifications. However both methods do not address the fault masking which could occur because of the analog stimulus circuit.

Another approach which has received favorable reviews is [10]. This method is designed to identify a set of basis vectors which span the space of possible device outputs. This acts to test the transfer function of the ADC, or more directly the INL and DNL. This method was conceived to save time on a mixed-signal tester. It is not primarily proposed as a BIST technique, but it may prove to be useful for BIST if it is combined with a suitable stimulus. Turning to the literature once again, we note that many authors recommend structural testing for BIST of mixed-signal ICs. A good example of this approach is proposed in [11], where the authors use one portion of the circuit as a reference to test another portion. While this approach will keep the area overhead low, there is an ambiguity in that if the IC "passes" the proposed BIST, does that mean all circuitry is indeed correctly operating within specifications, or can it mean that both the stimulus generator and the Circuit Under Test (CUT) are both out of spec yet each error cancels the other? That is to say, this approach is prone to the above-mentioned fault "masking" or having one fault being made unobservable by a second fault. Some authors may argue that this phenomenon is unlikely. However, in a high-volume commercially-produced product, any undesirable phenomenon which has even a remote probability of occurring will eventually manifest itself in the form of unhappy customers. Manufacturers are driven by market demands for defect levels numbering in only a few parts per million. In the final analysis, the BIST has to unambiguously assure that the circuit will meet any quality "assurance test" to which it may be subjected, or else the time and effort of implementing the BIST is simply a waste of money and resources.

Mixed-signal BIST may also prove useful for characterization or for diagnosis of field-returns. However, for these purposes, the location of the problem is of interest. There have been numerous suggestions to employ a "fault dictionary" approach to analog testing [12]. A fault dictionary is a list of all possible faults in the circuit such as the effect of open circuits, short circuits, or mismatched components on the output of the circuit. Fault dictionaries may work well to detect catastrophic failures, but they rapidly become unwieldy for detecting soft failures such as low gain. One serious disadvantage of a fault-dictionary approach is the fact that access to internal nodes of the circuit under test is usually required. This is usually impractical for sensitive nodes or nodes with very low-level signals. A second consideration is that for even a minor change in the structure of the circuit, the entire process of deriving a new fault dictionary must be repeated. It is possible that the method shown in [13] can assist in selecting the most important failure modes. Even then, the computational effort is considerable, and it is usually not economical to test for all possible faults. The approach is thus not suitable for manufacturing test. Furthermore, there is still the need for a precision analog stimulus.

Some authors propose that increasing the observability of nodes internal to the IC will facilitate BIST. This can be done by adding "probing" circuitry which will route the signals on the internal nodes out to pins of the IC. Among the several proposals for a scheme which will allow this kind of probing without increasing the pin count significantly, we will discuss two. The first example, which is targeted at Switched-Capacitor (SC) filters, is given in [14]. The authors propose reconfiguring the filter by means of the switches to change some of the SC filter stages to unity-gain buffers. This would allow an input signal to be passed to a particular filter stage, and permit the output response of the filter stage to be propagated to an output pin of the IC. Each stage of the filter could then be isolated for individual testing. A second proposal is found in [15] wherein the authors propose a system of dedicated unity-gain buffers and switches to inject signals into circuit nodes and carry the output response to pins. There are three issues which we feel would limit the use of these two approaches for BIST. First of all, there is again the possibility of a gain error in the circuit being masked by a second gain error in the on-chip probing or buffering circuits. Second, the connection of the probing, switches, or buffering circuitry to sensitive nodes of the circuitry on the IC can degrade the performance. Finally, there is still the need for a precise analog stimulus and a calibrated measurement instrument for the output response of the circuit under test. It would seem that these proposals will have somewhat limited application for an on-chip BIST per se, but there may be some application for a board or system-level BIST.

In conclusion, the BIST approach located in the literature for a mixed-signal IC that is closest to what industry is looking for is to be found in the AT&T patent [4] and recent contributions [8, 9] similarly attempting to obtain measures related to specifications. Manufacturers are wary of loop-back schemes due to the possibility of two faults "masking" each other. Other published approaches may be useful for board-level or system-level BIST, but none can perform dynamic testing such as FR, SNR, GT, IMD, or HD on-chip. This, therefore, is the challenge that will be described in the rest of this chapter. We shall first begin by describing several methods for on-chip signal generation.

6.3 On-Chip Signal Generation

In general, a test problem may be broken down into two aspects. First, the device-under-test (DUT) must be stimulated and then the output may be analyzed. This section will address the first task while the following section will deal with measurements. For spectral-based metrics, the stimulus should be a sinewave. Many analog oscillator designs are available for on-chip inclusion. However, the sinewave characteristics at the output will vary widely according to process variations. A calibration using off-chip analog instruments is necessary and it offsets the gains of BIST. It can be concluded that the dependency on analog components should be reduced to a minimum. Three different analog signal generation methods will be presented which make use of digital circuits. They differ in their hardware requirements, signal quality, and flexibility.

6.3.1 Digital Frequency Synthesis

Digital frequency synthesizers (DFS) generate analog signals using a cascade of a digital signal source and a digital-to-analog converter (DAC) [16]. A basic DFS architecture is shown in Figure 6-3. An integrator increments the phase variable, $\phi(n)$, by a frequency-control constant F. This value is used to reference a ROM which contains sine samples. The output is a digital sinewave which is converted to an analog signal via the DAC. Here the DAC is realized as a combination of a $\Delta\Sigma$ modulator followed by a 1-bit DAC (inverter circuit) and a lowpass (LP) filter circuit. This combination of circuitry is quite popular today for realizing high-resolution data converters with fine-lined VLSI processes. In fact, this architecture will be used quite often in this chapter. It will be demonstrated later that the LP filter needs not be explicit but may be part of the DUT.

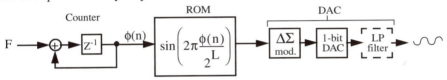

Figure 6-3 *Direct digital frequency synthesis.*

The frequency resolution of this device is a function of the number of samples stored in the ROM while the signal quality is limited either by the word length of each sample or the characteristics of the DAC. This circuit may easily be adapted to generate angle modulated communication signals. For example, frequency modulation is implemented by modifying the frequency-control word F while phase modulation involves incrementing the phase counter. The drawback of the ROM-based DFS is that the silicon area required is usually too large for analog and MS BIST. However, should some RAM be available during the analog test phase then it could be summoned for signal generation purposes.

6.3.2 Delta-Sigma Oscillators

An alternative to ROM-based synthesizers are the delta-sigma ($\Delta\Sigma$) oscillators [17]. This class of circuit is based on digital resonators where frequency control is implemented with a multiplication operation. However an N-bit by N-bit digital multiplier is slow when compared to adders and it occupies a large area on a die. A fixed-length multiplication is illustrated in Figure 6-4(a) with a coefficient c and a time-varying signal B. If the scalar can be made a power-of-two or a sum of a few powers-of-two, then the use of multipliers can be avoided. In many applications however, this restriction is unbearable as the location of circuit poles and zeros can be very sensitive to such quantization.

6.3.2.1   Delta-Sigma attenuator

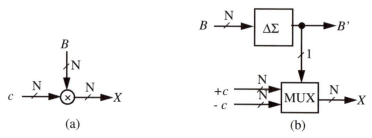

Figure 6-4 *(a) Multiplication; (b) Delta-sigma attenuator. (Figure courtesy of John Wiley & Sons, © 1997 [17].)*

The one-bit encoding property of $\Delta\Sigma$ modulators [18] can be used to simplify the multiplier, reducing it to an N-bit by 1-bit multiplication. This operation can then be implemented with a 2-to-1 multiplexer as shown in Figure 6-4(b). A necessary condition for this combination to emulate a multiplier, dubbed a $\Delta\Sigma$ attenuator [19], is that the $\Delta\Sigma$ modulator must have a unity signal transfer function (STF). In other words, the gain must be unity and have 0° phase shift, at least in the signal band. While this is not a property of most $\Delta\Sigma$ modulators, there are topologies that will guarantee this without constraining the designer in its choice of bandwidth or signal-to-noise ratio (SNR) [20]. Furthermore, the time-varying signal (B) is now available in 1-bit form (B') at the output of the $\Delta\Sigma$ modulator. The original signal will then be surrounded by out-of-band quantization noise which may later be eliminated by filtering. As explained in the section on DFS, the 1-bit representation of a signal is easier to convert to the analog domain as 1-bit DACs are compact and show excellent linearity. However, care must be exercised when placing a $\Delta\Sigma$ attenuator in a feedback loop as quantization noise is now introduced into the system. Therefore, coefficient c in Figure 6-4(b) should be kept as small as possible and much smaller than one in order to maintain the injected noise power to low levels.

6.3.2.2   Low-pass Delta-Sigma Oscillator

The original circuit making use of a $\Delta\Sigma$ modulator in the feedback loop of a digital resonator, now labeled a low-pass $\Delta\Sigma$ oscillator, was introduced by Lu *et al.* [21].

The device, shown in Figure 6-5 is made of two digital integrators, a low-pass $\Delta\Sigma$ modulator and a multiplexer. The output of the $\Delta\Sigma$ modulator has been labeled $X(z)$ for reference. The $\Delta\Sigma$ modulator encodes its input on a single-bit stream and the multiplexer realizes a 1-bit by N-bit multiplication. For this operation to emulate a digital multiplier, the LP $\Delta\Sigma$ modulator must have a unity signal transfer function (STF). The circuit may also accommodate an LP $\Delta\Sigma$ modulator that has an STF equal to z^{-1} (such as the classical second-order LP $\Delta\Sigma$ modulator [22] shown in Figure 6-6) but the resonator must be modified by trading the delayed integrator for a non-delayed one.

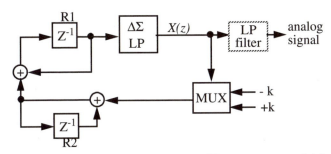

Figure 6-5 *Low-pass delta-sigma oscillator. (Figure courtesy of John Wiley & Sons, © 1997 [17].)*

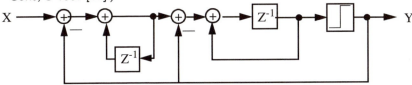

Figure 6-6 *Second-order low-pass delta-sigma modulator. (Figure courtesy of John Wiley & Sons, © 1997 [17].)*

From an analysis of the circuit shown in Figure 6-5, the characteristic equation is found to be:

$$z^{-2} - (2 - k)z^{-1} + 1 = 0 \tag{6.1}$$

With $0 \le k \le 4$, the poles lie on the unit circle of the z-plane resulting in oscillation. Using Eq. (6.1), the frequency of oscillation (F_o) can be related to k and the clock frequency (F_s) according to

$$\Omega_o = \frac{2\pi F_o}{F_S} = \mathrm{acos}\left(1 - \frac{k}{2}\right) \tag{6.2}$$

where Ω_o is the normalized angular frequency expressed in radians. A frequency analysis reveals that the 1-bit circuit output contains a very high-quality low-frequency analog sinewave and high-frequency noise. The analog sinewave signal can then be isolated by low-pass filtering. The amplitude and phase of the sinewave will depend on the initial conditions of R1 and R2, denoted ϕ_1 and ϕ_2, respectively. Modeling the $\Delta\Sigma$ modulator and multiplexer combination as a perfect multiplier and using simple z-domain arithmetic, the following equation for the signal at the output of the $\Delta\Sigma$ modulator is obtained:

$$X(z) = \frac{(1 - z^{-1})\phi_1 + z^{-1}\phi_2}{1 - (2 - k)z^{-1} + z^{-2}} \tag{6.3}$$

Using the inverse z-transform, a more informative time-domain signal description is obtained:

$$x(nT) = \phi_1 \frac{\sqrt{2(1 - \cos\Omega_o)}}{\sin\Omega_o} \sin\left(\Omega_o nT + \frac{\pi + \Omega_o}{2}\right) + \phi_2 \frac{1}{\sin\Omega_o} \sin(\Omega_o nT) \quad (6.4)$$

where T is the clock period and n is a time index. The first term is the oscillation due to the initial condition of register 1 while the second term describes the sinewave triggered by the initial value of register 2. The amplitude and phase of the sinusoidal signal can be selected separately since the two terms are linearly independent.

The quality of the generated signal was assessed through prototyping with a Xilinx XC4010 FPGA. Figure 6-7 (a) shows the power density spectrum of a 5 kHz signal up to the 1 MHz clock frequency obtained from a spectrum analyzer with the reference impedance set to 50 ohms. The signal band, from DC to 10 kHz, is displayed on a larger scale in Figure 6-7 (b). There is very little phase noise around the signal and its power is 80 dB above the noise floor.

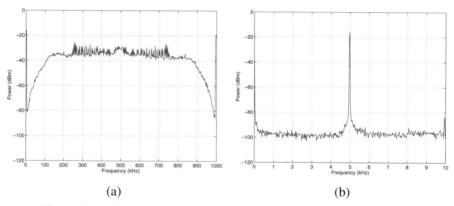

(a)   (b)

Figure 6-7 *Low-pass ΔΣ oscillator measured power density spectrum: (a) Up to clock frequency; (b) Signal band. (Figure courtesy of John Wiley & Sons, © 1997 [17].)*

6.3.2.3   Bandpass Delta-Sigma Oscillator

The basic ΔΣ oscillator was later modified for bandpass (BP) ΔΣ modulation [23]. The resulting circuit, shown in Figure 6-8, may have the signal band located at any ratio of the clock frequency [24]. The network is composed of a second-order resonator and a feedback loop composed of a bandpass ΔΣ modulator in series with a 2-to-1 multiplexer. The output of the ΔΣ modulator is denoted $X(z)$. Coefficient K_c is a coarse tuner as it sets the center of the oscillation frequency range. Fine tuning the oscillation around this center frequency is performed with coefficient k_f. For a smaller silicon area, K_c should be a power-of-two or a sum of a small number of power-of-two terms to avoid the need for a full-fledged multiplier. In particular, the use of $K_C = 0$ results in the signal band being located at a quarter of the clock frequency and economical BP ΔΣ modulator implementations.

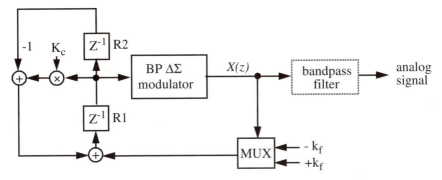

Figure 6-8 *Bandpass delta-sigma oscillator. (Figure courtesy of John Wiley & Sons, © 1997 [17].)*

The 1-bit signal at the output of the $\Delta\Sigma$ modulator contains the oscillatory signal plus out-of-band quantization noise. The sinewave is extracted by bandpass filtering this bit stream. It should be noted again that the $\Delta\Sigma$ modulator must have a unity signal transfer function (STF). The characteristic equation of the bandpass $\Delta\Sigma$ oscillator is

$$z^{-2} + (k_f - K_c)z^{-1} + 1 = 0 \tag{6.5}$$

from which the relation governing the frequency of oscillation can be obtained as follows:

$$\Omega_o = \frac{2\pi F_o}{F_S} = \mathrm{acos}\left(\frac{K_c - k_f}{2}\right) \tag{6.6}$$

The amplitude and phase of the sinewave depend on the initial value of register R1 and R2, once again, denoted ϕ_1 and ϕ_2, respectively. Using simple analysis, one can write:

$$X(z) = \frac{\phi_1 - \phi_2 z^{-1}}{1 + (K_C - k_f)z^{-1} + z^{-2}} \tag{6.7}$$

from which the corresponding time-domain equation is found to be:

$$x(nT) = \phi_1 \frac{1}{\sin\Omega_o} \sin(\Omega_o nT + \Omega_o) + \phi_2 \frac{1}{\sin\Omega_o} \sin(\Omega_o nT) \tag{6.8}$$

Because the two sinusoids have a Ω_o phase difference, the amplitude and phase of the signal can be selected independently. Like before, if the exact value of phase is not a concern, we can set $\phi_2 = 0$ and select the desired amplitude through ϕ_1.

The stability of the device as well as the signal quality were verified experimentally. The $\Delta\Sigma$ modulator used in this experiment is shown in Figure 6-9. It has unity STF, fourth-order noise shaping, and does not require multipliers as its coefficients are powers of two or a sum of two powers-of-two. The signal band is centered at $\pi/2$ and the oversampling ratio (OSR) is 25. It should be clear that the OSR only refers here to the ratio of the bandwidth to the Nyquist frequency as a single clock is used for the whole network.

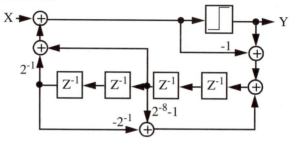

Figure 6-9 *A fourth-order bandpass delta-sigma modulator. (Figure courtesy of John Wiley & Sons, © 1997 [17].)*

The noise transfer function ($H(z)$) of this $\Delta\Sigma$ modulator is

$$H(z) = \frac{z^{-4} + (2 - 2^{-8})z^{-2} + 1}{2^{-1}z^{-4} + (1 - 2^{-8})z^{-1} + 1} \tag{6.9}$$

The circuit is implemented on an FPGA clocked at 1 MHz for fast prototyping. With $K_C = 0$, the feedback coefficient is chosen to be 0.0628 triggering a frequency of oscillation of 255 kHz. The initial value of the registers are 0.5 and 0.0 for R1 and R2, respectively, yielding an amplitude of oscillation of 0.5 (signal power 9 dB lower than the bit-stream power). The output is fed to a 1-bit DAC that converts the digital sequence to a polar-return-to-zero coded signal. The circuit output was validated using a spectrum analyzer. Figure 6-10 (a) shows the power density spectrum in the Nyquist interval (0 to 500 kHz) while part (b) displays the signal band (240 to 260 kHz) in more detail.

As expected the sinusoidal signal is located at 255 kHz and the quantization noise is pushed out of the region surrounding the tone. The measured signal-to-noise ratio (SNR) is 50.9 dB for an OSR of 25 (20 kHz bandwidth). Transmission zeros are visible at 245 kHz and 250 kHz. The oscillator was left running for over 12 hours with no change in the output signal, indicating long-term stability. This signal was also filtered by a fourth-order bandpass filter. The result is shown in Figure 6-11 (b). The frequency of the sinusoid was measured to be exactly 255 kHz. To illustrate the relationship between the generated sinusoid and the clock frequency, the bit stream output is shown in Figure 6-11 (a). Approximately 4 pulses of the sequence make up a single period of the sinusoidal signal, thus illustrating the capability of this circuit to generate sinusoids at high ratios of the clock frequency.

6.3.2.4   *Multi-Tone Delta-Sigma Oscillators*

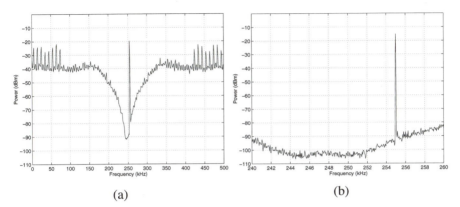

Figure 6-10   *Bandpass ΔΣ oscillator measured power-density spectrum: (a) Nyquist interval; (b) Signal band. (Figure courtesy of John Wiley & Sons, © 1997 [17].)*

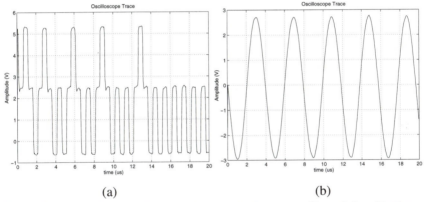

Figure 6-11   *(a) Signal in the time domain; (b) Bandpass filtered signal in the time domain. (Figure courtesy of John Wiley & Sons, © 1997 [17].)*

Multi-tone signals are essential for the measurement of analog circuit parameters such as inter-modulation distortion. The basic design was therefore modified to allow multi-tone signal generation. A hardware efficient technique to realize this has been demonstrated in [25]. Time division multiplexing is used and the resulting bit streams are interleaved at the output producing a single bit stream. The price to pay, however, is a reduction in the effective clock frequency by a factor equal to the number of tones. Figure 6-12 shows the circuit of Figure 6-5 modified for two-tone signal generation using time-division multiplexing. The combinational part of the circuit remains the same except that the 2:1 multiplexer is exchanged for a 4:1 multiplexer. All registers have been replaced by strings of two registers. However, on any clock cycle only the content of the first of these two registers has an effect on the combinational circuit. The content of the second one is just passed on to the first to be used in the following clock cycle.

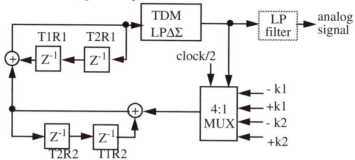

Figure 6-12 *Two-tone low-pass delta-sigma oscillator.*

Figure 6-13 shows the second-order LP $\Delta\Sigma$ modulator of Figure 6-6 adapted to handle a two-tone time-division multiplexed input signal.

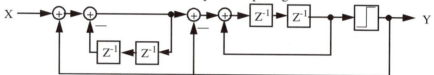

Figure 6-13 *Time-division multiplexed second-order low-pass delta-sigma modulator.*

Just like the LP $\Delta\Sigma$ oscillator, the BP $\Delta\Sigma$ oscillator may be economically adapted for multi-tone generation using time-division multiplexing. Figure 6-14 shows the resulting circuit when this technique is applied to the circuit of Figure 6-8.

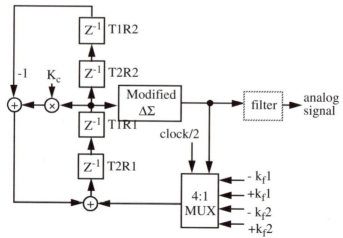

Figure 6-14 *Two-tone bandpass delta-sigma oscillator.*

The two-tone ΔΣ oscillator of Figure 6-14 was implemented on a single XC4010 FPGA. The circuit uses 74% of the function generator resources of a Xilinx XC4010 and 42% of the flip-flops. Figure 6-15 (a) shows the spectral density up to half the 1 MHz system clock. Because time-multiplexing is used, the effective clock is reduced to 500 kHz and two images are visible. Figure 6-15 (b) shows the first signal band. The signals appear at 124 kHz and 127.5 kHz. The spurious tones at 120 kHz and 132 kHz were not present in the simulation and are probably caused by the non-ideal behavior of the 1-bit DAC. The tones exhibit different powers because of second-order effects of the ΔΣ modulator [26]. Another method may also be used to add the bit stream and thus operate at the full clock frequency but it requires more hardware [27].

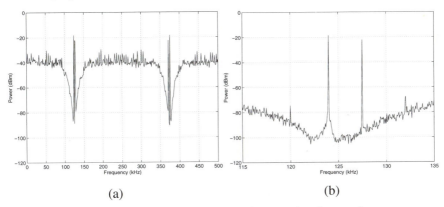

Figure 6-15 *Two-tone measured power density spectrum: (a) Nyquist interval; (b) Signal band.*

6.3.2.5   FM Signal Generator

The first two ΔΣ oscillators are capable of generating only sinusoidal signals. While a pure sinewave may verify many specifications, it is insufficient for complex communication circuits. For example, wireless transceivers use modulation schemes to encode data and therefore the circuitry involved in demodulation sees a steady signal and no time-varying information. One of these modulation schemes, frequency modulation (FM), is the preferred one for analog wireless communication. A frequency modulated signal generator was thus designed by cascading the low-pass and bandpass ΔΣ oscillators [28]. The low-pass ΔΣ oscillator generates a low frequency digital sinewave which modulates the frequency of oscillation of the bandpass ΔΣ oscillator as illustrated in Figure 6-16. Scaling by a factor S is used to adjust the frequency deviation. This constant should be a power of two as fine tuning of the sinewave amplitude at the input of the bandpass ΔΣ oscillator can be done by modifying the initial conditions of the registers of the lowpass ΔΣ oscillator. It is interesting to note that the input of the multiplexer in the $\Delta\Sigma$ oscillator is no longer a constant as in previous designs but a time-varying signal. The $\Delta\Sigma$ attenuator concept is therefore not restricted to scaling operations. The bandpass $\Delta\Sigma$ oscillator can thus be viewed as a digital implementation of the well-known voltage controlled oscillator (VCO) circuit.

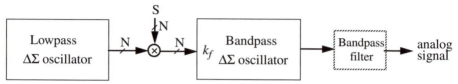

Figure 6-16 *Frequency modulated signal generator based on $\Delta\Sigma$ oscillators. (Figure courtesy of John Wiley & Sons, © 1997 [17].)*

While the carrier frequency is set to a quarter of the clock frequency, a wide range of modulated signal frequencies can be obtained as this is set by the oscillating frequency of the low-pass $\Delta\Sigma$ oscillator (Eq. (6.2)). The modulation index (β) of the FM signal will depend on the maximum value at the input of the bandpass $\Delta\Sigma$ oscillator (k_{fm}) and the low-pass $\Delta\Sigma$ oscillator feedback coefficient according to:

$$\beta = \frac{k_{fm}}{2\,\text{acos}\left(1 - \frac{k}{2}\right)} \tag{6.10}$$

Again the circuit was implemented on an FPGA for fast prototyping and the measured power density spectrum is shown in Figure 6-17 for both the Nyquist interval and the signal band. The circuit was clocked at 1.818 MHz and therefore the carrier was located at a quarter of this frequency (454.5 kHz). A modulated signal frequency of 2 kHz with a modulation index equal to 3 was selected. In Figure 6-17 (b), sidebands characteristic of FM modulation appear at 2 kHz intervals from the 454.5 kHz carrier frequency. The signal was then demodulated by a series combination of a bandpass filter, a limiter, and a Foster-Seeley detector.

Figure 6-18 (a) shows the resulting signal as seen on an oscilloscope. The quality of the signal may be further assessed by examining the power density spectrum of this signal shown in Figure 6-18 (b).

6.3.3   Fixed-Length Periodic Bit Stream

The third scheme is more efficient in terms of hardware but generally results in lower signal quality. The hardware overhead could in fact be negligible as available digital structures such as boundary scan chains or on-chip RAM can be used [29]. The principle, illustrated in Figure 6-19, is to repeat a fixed-length bit stream thus making it periodic. To find this fixed-length bit stream, an aperiodic one-bit sig-

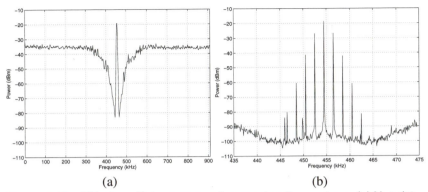

Figure 6-17 *FM ΔΣ oscillator measured power density spectrum: (a) Nyquist interval; (b) Signal band. (Figure courtesy of John Wiley & Sons, © 1997 [17].)*

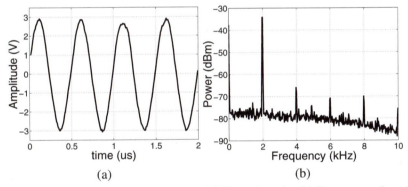

Figure 6-18 *Demodulated FM signal: (a) Time domain; (b) Frequency domain.*

nal is first generated by a software sinewave generator and ΔΣ modulator and then a subset of this bit stream is selected. After storing it in memory, the subset is repeated. Through this method, any periodic signal can be represented with a very small number of bits, on the order of a hundred. However, a different bit stream is required for each frequency and amplitude such that, in contrast to the fixed size of a DFS ROM, the memory requirement increases with the number of signals. The method may in fact be seen as a special case of DFS [30].

To illustrate the quality of the signals that can be obtained with this method, the power density spectrum of a repeated 392-bit stream as read on a spectrum analyzer is shown in Figure 6-20 (a). Here a 10 kHz signal is generated from a 1.3 MHz clock. In Figure 6-20 (b), the output was sent to a fourth-order low-pass filter with cut-off frequency at 10 kHz and viewed on an oscilloscope. Other types of signals can also be generated with the same hardware setup such as different amplitudes, phases, or shapes.

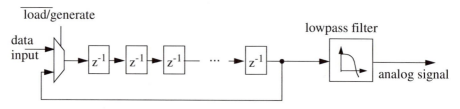

Figure 6-19   *Fixed-length periodic bit stream signal generation.*

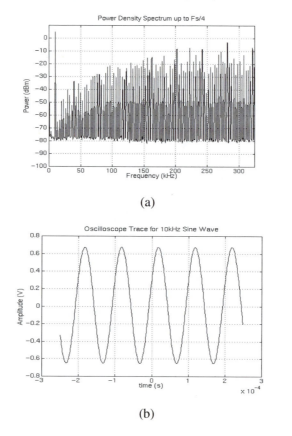

(a)

(b)

Figure 6-20   *Fixed-length periodic bit stream measured power-density spectrum: (a) Nyquist interval; (b) After filtering. (Figure courtesy of the IEEE, © 1996 [29].)*

6.4   Parameter Analysis

As many analog specifications are frequency-based, a way of separating sinusoids in the output signal as well as sinusoid from noise is required. Three methods are available for this task. The industry standard is the Fast Fourier Transform (FFT) but it requires a large amount of resources and is not always suitable for on-chip implementation. If the number of frequency components of interest is small then a sinewave fitting algorithm may be used instead. An alternative method which uses less resources is digital filtering.

6.4.1   Fast Fourier Transform

The FFT is a tool for decomposing a finite length signal into a set of harmonically related frequency components. The sample period should ideally be a multiple of the period of each sinewave in the stimulus [31]. Separation of signal and noise using an N-point FFT yields very accurate results. However, it necessitates N memory cells to store the samples and requires on the order of N·logN operations. Since measurement accuracy increases with the number of samples, in many systems the overhead to implement the FFT extraction procedure may be prohibitive.

6.4.2   Sinewave Correlation

If the spectral power at only one frequency is of interest, then a correlation-type digital power meter can be used. A block diagram of this method is shown in Figure 6-21. The input signal is multiplied by the in-phase and quadrature component of a sinewave. The powers of the resulting product signals are computed and then added. This method is less computationally expensive than the FFT if only a small number of frequencies are of interest. However it requires the sine and cosine functions, either computed directly or stored in a look-up ROM. An IEEE standard is devoted to this technique [32].

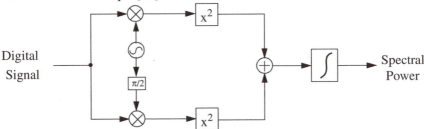

Figure 6-21   *Measuring spectral power with correlator.*

6.4.3   Bandpass Filters

The power of sinewave components can also be determined using a combination of digital filters, squaring functions, and integrators. To reduce hardware requirements, a circuit implementing both the bandpass function and the band-reject function was used in [33] and [34]. It is based on a time-recursive implementation of an arbitrary transform [35]. The signal flow graph for a filter with N bandpass outputs is shown in Figure 6-22. The input is identified with the variable X. The bandpass outputs are labeled Y_n while E denotes the band-reject output. The coefficients k_n set the frequency of each of the bandpass sections while k_{BW} primarily determines the pole radius. Signal power is computed by squaring and adding the output of the corresponding bandpass filter while the noise power is obtained by processing the band-reject output in a similar manner.

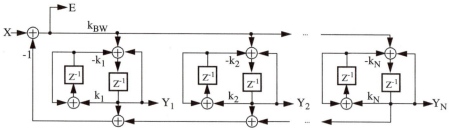

Figure 6-22 *Narrowband digital filter.*

While the second-order filter presented here is very efficient, it suffers from an extremely narrow notch function. However, a more tolerant extraction circuit may be obtained using higher-order filters. The accuracy obtained with digital filters depends on the selectivity of the filtering function and on the number of samples. The selectivity is in turn a function of the position of the roots with respect to the unit circle in the z-plane. Roots near the unit circle will yield accurate measures but will require long conversion times. The effect of the number of samples on accuracy is illustrated in Figure 6-23 where each circle represents a simulation example. The spread of the estimated variable decreases with increasing sample size.

The choice of FFT, sinewave fitting algorithm or digital filters to implement parameter extraction depends on the availability of computing resources in the system and the required accuracy.

6.5 Application: MADBIST

Two BIST schemes, labeled mixed analog-digital built-in self-test (MADBIST), have been proposed. The first one is for baseband devices and has been demonstrated for a voice-band CODEC [33]. The second one targets bandpass devices such as wireless transceivers [36].

6.5.1 Baseband MADBIST

It was argued at the beginning of this chapter that testing two blocks in series may cause error masking. Therefore in MADBIST, to test devices such as the one

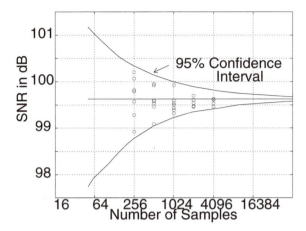

Figure 6-23 *Convergence of SNR with sample size.*
(Figure courtesy of the IEEE, © 1995 [33].)

illustrated in Figure 6-1, the ADC is first verified alone before closing the loop. To implement the BIST capability, a low-pass ΔΣ oscillator and a multiplexer have been added. A diagram of the test setup is shown in Figure 6-24. A pulse density modulated (PDM) signal from the low-pass ΔΣ oscillator is applied to the input of the ADC. In the PDM bit stream is a high-quality low-frequency sinewave and high-frequency quantization noise. This noise is removed by the anti-aliasing filter (AAF) leaving a spectrally-pure sinusoid to excite the remaining circuitry. The digital output of the ADC is then analyzed by the DSP to separate the signal and the noise using an FFT or a digital filter. Analog measures such as signal-to-noise ratio (SNR), gain tracking, and frequency response may then be obtained. Since the low-pass ΔΣ oscillator may further be adapted to provide multi-tone signals, intermodulation distortion measurements [34] and rapid frequency response tests are also possible.

After the ADC has passed all the tests, the DAC may now be verified by placing it in the loop as shown in Figure 6-25. A digital signal, generated by the ΔΣ oscillator or the DSP, excites the DAC. The analog output is sent to the ADC which converts it back into digital form. A digital filter or FFT implemented by the DSP may then separate the noise from the signal.

After the DAC is characterized, other analog circuits may be tested by placing them between the DAC and the ADC as illustrated in Figure 6-26.

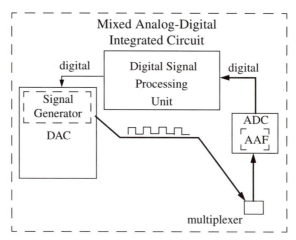

Figure 6-24 *Analog-to-digital converter test setup.*

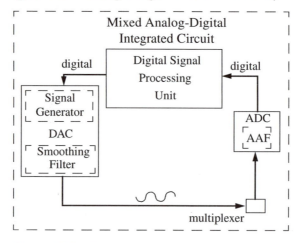

Figure 6-25 *Digital-to-analog converter test setup.*

6.5.2   Baseband MADBIST Experiments

A block diagram of the experimental setup is shown in Figure 6-27 (a). A low-pass $\Delta\Sigma$ oscillator stimulated the device under test and the output is analyzed using digital filters. The device under test is an ADC chip supplied by Nortel. It includes a second-order $\Delta\Sigma$ modulator operated at 2.048 MHz with a bandwidth of 3.3 kHz. Results obtained will be compared with measures from a second apparatus, illustrated in Figure 6-27 (b), which relies on a commercial signal source for stimulus generation and the FFT for parameter extraction.

The first experiment was the measure of the SNR as the input signal is increased. The $\Delta\Sigma$ oscillator and a commercial signal source provided the 1031.25

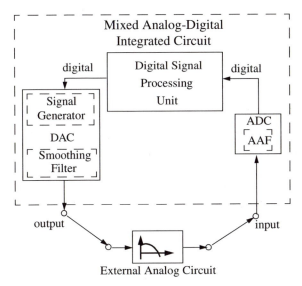

Figure 6-26 *External circuit test setup.*

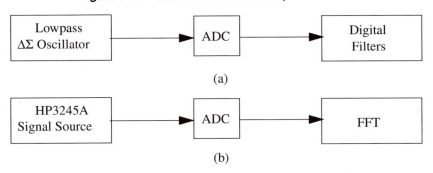

Figure 6-27 *(a) Baseband MADBIST experiment; (b) Characterization.*

Hz stimulus. The results of the experiment are shown in Figure 6-28. Both methods yield comparable results.

The second experiment measured the relative gain of the device under test by computing the output power and comparing it with the input. The test frequency was again selected to be 1031.25 Hz. The results are plotted in Figure 6-29 and it can be seen that the MADBIST schemes compares very well with the reference method.

External circuits have also been tested [37]. Results are displayed in Figure 6-30 for the frequency response of an analog second-order bandpass filter using the arrangement shown in Figure 6-26. Again the validity of the new measuring method is demonstrated as both results sets match closely.

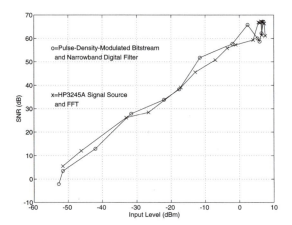

Figure 6-28 *SNR versus input signal level. (Figure courtesy of the IEEE, © 1995 [33].)*

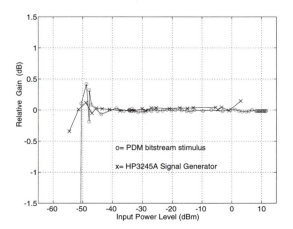

Figure 6-29 *Gain tracking. (Figure courtesy of the IEEE, © 1995 [33].)*

6.5.3   MADBIST for Transceiver Circuits

The baseband MADBIST scheme was later extended to bandpass devices such as those found in wireless transceivers. A block diagram of a wireless transceiver is shown in Figure 6-31. On the left side are the RF and IF stages mostly composed of analog circuits. The baseband section is made of a voice-band codec, a digital signal

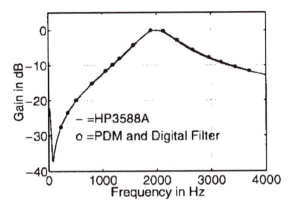

Figure 6-30 *Frequency response of a second-order bandpass filter.*

processor (DSP), and an RF codec which modulate/demodulates low-frequency signals coming from the analog front-end. The MADBIST scheme may be applied to the audio codec and one can also adapt it to most of the remaining analog circuitry, the IF stage, and the RF codec, using the bandpass $\Delta\Sigma$ oscillator. The only change in the circuit that is required to make it self-testable is the addition of two multiplexers at the input of the receiver RF and IF stages shown in Figure 6-31.

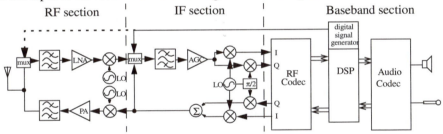

Figure 6-31 *Wireless communication system.*

The diagram for the receiver test setup is shown in Figure 6-32. The test signal path is illustrated using bold lines. The BP $\Delta\Sigma$ oscillator feeds the high-frequency analog signal imbedded in a 1-bit stream to the IF stage. The out-of-band noise is removed by the IF filter. The RF codec demodulates the down-converted signal and the DSP unit then analyzes the output with one of the methods outlined in Section 6.4 and decides if the device meets the necessary specifications.

After validation of the receiving side, the transmitter IF section may now be tested. A diagram of the test setup is shown in Figure 6-33. A digital signal, either from the BP $\Delta\Sigma$ oscillator or from the DSP, excites the transmitter which generates an IF signal. This IF signal is then re-routed and applied to the receiver path via the multiplexer. The receiver down-converts and demodulates the output to a voice-band

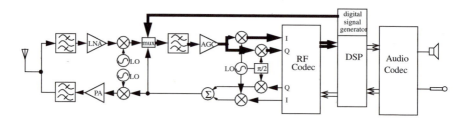

Figure 6-32 *Wireless receiver test setup.*

digital signal. The DSP may then analyze the result and either accept or reject the circuit.

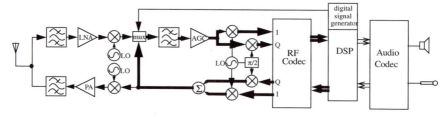

Figure 6-33 *Wireless transmitter test setup.*

The test methodology described here can be repeated for the RF stage of the receiver. However, practically speaking, creating high quality test signals in the 0.8 to 2 GHz range is no easy task and further research is necessary to determine whether it is possible to implement.

6.5.3.1   Experimental Results

To illustrate the feasibility of this scheme, the experimental setup of Figure 6-34 was built. A BP $\Delta\Sigma$ oscillator is implemented using an XC4010 FPGA. The stimulus is passed through a BP filter to simulate the frequency selectivity of wireless receivers. The filtered signal is then down-converted by a balanced modulator with the second input being a pure sinusoid. The output is filtered to remove the second modulation product and then sampled at 40 kHz by an HP 1430A ADC. The carrier frequency (F_c) was selected to be 448 kHz while the modulated signal frequency (F_m) was set to 7 kHz. The $\Delta\Sigma$ oscillator thus generated a 455 kHz sinewave while being clocked at 1818 kHz. An FFT of the baseband sampled signal is shown in Figure 6-35 (a). A raised cosine window was applied to the samples to eliminate leakage effects caused by non-coherent sampling. If the filtering scheme is used, then the application of a window on the samples prior to signal extraction is not necessary. Returning to the results displayed in Figure 6-35 (a), we see that a tone is present at 7 kHz as expected. For comparison, the experiment was repeated with a sinewave from an HP 3314A signal generator. The spectral content of the sampled baseband signal is shown in Figure 6-35 (b). The results are very similar to those in

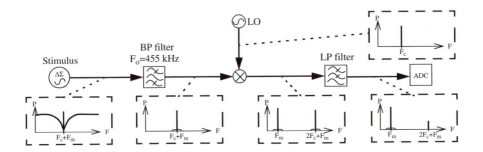

Figure 6-34 *Intermediate frequency MADBIST experimental setup.*

(a) except for some spurious tones that are a result of our noisy experimental environment.

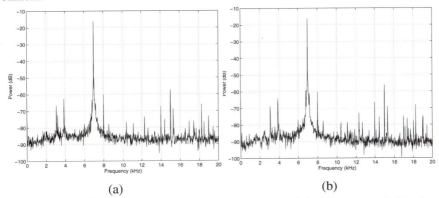

(a)   (b)

Figure 6-35 *Power density spectrum of sampled down-converted signal: (a) Delta-sigma oscillator; (b) HP 3314A signal source.*

Since the validity of the BP $\Delta\Sigma$ oscillator as a test stimulus for the wireless receiver has been established, various tests were performed to obtain meaningful analog metrics.

6.5.3.2   Signal-to-Noise Ratio

This experiment was performed with both the BP $\Delta\Sigma$ oscillator and a sinewave from an HP 3314A signal generator. The results are shown in Figure 6-36. The two sources yield similar measures especially at low signal levels. At the high end, the discrepancy between the two curves illustrates the limitations of the filter method for separation of signal and noise. Indeed in this experiment a significant amount of noise is located at frequencies close to the signal frequency. Since a brick-wall filter is not possible, the band-reject filter, while eliminating the signal, will attenuate this noise, thus causing the observed increase in the SNR measure.

6.5.3.3   Frequency Response

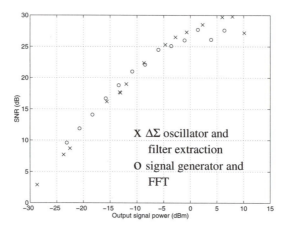

Figure 6-36 *SNR of the sampled down-converted signal with delta-sigma oscillator and HP 3314A signal generator. (Figure courtesy of the IEEE, © 1995 [36].)*

Again the experiment was done with the BP $\Delta\Sigma$ oscillator as well as a sine-wave from a signal generator for reference. On account of that the amplitude of the signal generated by the BP $\Delta\Sigma$ oscillator is not exactly governed by Eq. (6.8), it is difficult to find the initial values for the two registers that will maintain a constant amplitude for a wide range of test frequencies; variations in signal amplitude of 5 dB over a frequency range of 20 kHz have been observed. So to avoid a situation where the input signal is not what is expected, we first run a calibration cycle where we determine the signal amplitude received by the extraction circuitry and use these values to correct any future measurements. Fortunately, this entire process involves deterministic digital circuits and therefore can be performed off-chip using a digital simulator that is capable of accounting for finite-length register effects. The frequency response of the receiver is shown in Figure 6-37 for the two different types of excitation. Clearly, they correspond quite closely. Also seen is the frequency response behavior of the BP filter used in this experiment. This plot was included to provide a sense of the expected channel characteristics.

6.5.3.4   Intermodulation Distortion

A significant analog metric of narrowband systems, intermodulation distortion, may not be obtained using a single tone as stimulus. The two-tone BP $\Delta\Sigma$ oscillator of Figure 6-14 was therefore implemented for the following experiment. The circuit was clocked at 1212 kHz and the second signal band, centered around 454.5 kHz, was utilized for this test. By generating two sinewaves at 151 and 152 kHz, images at 454 and 455 kHz were created. Since the carrier frequency was 448 kHz, two sinusoids appeared in the baseband at 6 and 7 kHz. The power spectral density (PSD) of the sampled output is shown in Figure 6-38. Intermodulation distortion

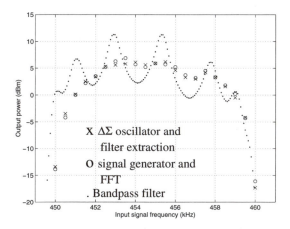

Figure 6-37 *Frequency response of receiver for delta-sigma oscillator and HP 3314A signal generator. (Figure courtesy of the IEEE, © 1995 [36].)*

products (F_2-F_1, $2F_1$-F_2,...) are clearly visible. The other signals are spurious tones introduced by our measurement set up. Similar results would be obtained with commercial waveform generators.

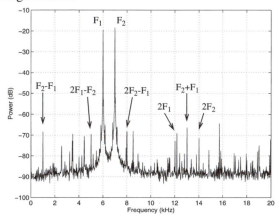

Figure 6-38 *Power spectral density of sampled down-converted two-tone delta-sigma oscillator signal. (Figure courtesy of the IEEE, © 1995 [36].)*

6.6 Conclusions and Future Directions

Mixed-signal BIST is beneficial in many ways. Not only does it decrease the test cost but it also adds value to the product in allowing it to be verified or tuned in the field. Furthermore, it reduces the loading problems introduced when the signals have to travel between the device-under-test and the tester. Although the benefits of BIST have been clearly established, IC manufacturers are still reluctant to implement it. An accurate cost analysis is probably required before they commit to the concept. The BIST schemes we propose are sinusoidal-based and thus can obtain measures relevant for many analog specifications. They rely on analog sources which are almost entirely digital. Since the analog dependency is reduced to a 1-bit DAC, calibration is not required. Our method requires the presence of some form of analog-to-digital conversion as the parameter extraction is done in the digital domain. A future research topic is the development of analog instruments with reduced analog dependencies so that we can extend BIST to any analog circuit.

References

1. V. D. Agrawal, C. R. Kime, and K. K. Saluja, "A tutorial on built-in self-test; part1: principles," *IEEE Design and Test of Computers*, vol. 10, no. 2, pp. 73-82, March 93.
2. L. Milor and A.L. Sangiovanni-Vincentelli, "Minimizing production test time to detect faults in analog circuits," *IEEE Trans. on Computer-Aided Design of Integrated Circuits and Systems*, vol. 13, no. 6, pp. 796-813, June 1994.
3. G. W. Roberts, "Metrics, techniques and recent developments in mixed-signal testing," *Proc. 1996 IEEE/ACM International Conference on Computer Aided Design*, pp. 514-521, San Jose, California, November 1996.
4. M. R. DeWitt, G. F. Gross, and R. Ramachandran, AT&T Bell Laboratories, Murray Hill, NJ, "Built-in self-test for analog to digital converters," United States Patent No. 5,132,685, filed Aug. 9, 1991, granted July 21, 1992.
5. M. J. Ohletz, "Hybrid built-in self-test (HBIST) for mixed analogue/digital integrated circuits," *Proc. 1991 European Test Conference*, pp. 307-316, Munich, Germany, April 1991.
6. F. F. Tsui, *LSI/VLSI Testability Design*, McGraw-Hill 1986.
7. P. H. Bardell, W. H. McAnney, and J. Savir, *Built-In Test for VLSI: Pseudorandom Techniques*, John Wiley & Sons Inc., New York 1987.
8. S. Sunter and N.Nagi, "A simplified polynomial-fitting algorithm for DAC and ADC BIST," *Proc. 1997 International Test Conference*, pp. 389-395, Washington DC, October 1997.
9. A. Frisch and T. Almy, "HABIST: histogram-based analog built-in self-test," *Proc. 1997 International Test Conference*, pp. 760-767, Washington DC, October 1997.
10. T. M. Souders and G. N. Stenbakken, "Comprehensive approach for modeling and testing analog and mixed-signal devices," *Proc. 1990 International Test Conference*, pp. 169-173, Washington DC, September 1990.
11. K. Arabi, B. Kaminska, and J. Rzeszut, "A new built-in self-test approach for digital-to-analog and analog-to-digital converters," *Proc. 1994 International Conference on CAD*, pp. 491-494, San Jose CA, November 1994.
12. E. P. Ratazzi, "Fault coverage measurement for analog circuits," *Proc. 1992 IEEE Custom Integrated Circuits Conference*, pp 17.2.1-17.2.4, 1992.
13. L. Milor and A. Sangiovanni-Vincentelli, "Optimal test set design for analog circuits," *Proc. 1990 IEEE International Conference on Computer Aided Design*, pp 294-297, 1990.
14. M. Soma and V. Kolarik, "A design-for-test technique for switched-capacitor filters," *Proc. 1994 IEEE VLSI Test Symposium*, pp. 42-47, Cherry Hill, NJ, April 1994.
15. C. L. Wey, "Built-in self-test (BIST) structure for analog circuit fault diagnosis," *IEEE Trans. on*

Instrumentation and Measurement, vol. IM-39, pp. 517 -5 21, June 1990.

16. J. Tierney, C.M. Rader, and B. Gold, "A digital frequency synthesizer," *IEEE Trans. on Audio Electroacoustic*, vol. 19, pp. 48-57, 1971.

17. B.R. Veillette and G.W. Roberts, "Delta-sigma oscillators: versatile building blocks," *International Journal of Circuit Theory and Applications*, vol. 25, pp. 407-418, 1997.

18. M.W. Hauser, "Principles of oversampling A/D conversion," *Journal of the Audio Engineering Society*, vol. 39, No. 1/2, pp.3-26, January/February 1991.

19. D.A Johns and D.M. Lewis, "Design and analysis of delta-sigma based IIR filters," *IEEE Trans. Circuits and Systems-II*, vol. 40, no. 4, pp. 233-240, April 1993.

20. X. Haurie and G. W. Roberts, "Arbitrary-precision signal generation for bandlimited mixed-signal testing," *Proc. 1995 IEEE International Test Conference*, Washington, pp. 78-86, October 1995.

21. A.K. Lu, G.W. Roberts, and D.A. Johns, "A high-quality analog oscillator using oversampling D/A conversion techniques," *IEEE Trans. Circuits and Systems-II*, vol. 41, no. 7, pp. 437-444, July 1994.

22. J.C. Candy, "A use of double integration in sigma-delta modulation," *IEEE Trans. Communications*, vol. 33, no. 3, pp. 249-258, March 1985.

23. R. Schreier and M. Snelgrove, "Bandpass sigma-delta modulation," *Electronics Letters*, vol. 25, no. 23, pp. 1560-1561, November 1989.

24. B.R. Veillette and G.W. Roberts, "High frequency sinusoidal generation using delta-sigma modulation techniques," *Proc. 1995 IEEE International Symposium on Circuits and Systems*, Seattle, pp. 637-640, May 1995.

25. A.K. Lu and G.W. Roberts, "An analog multi-tone signal generator for built-in self-test applications," *Proc. 1994 IEEE International Test Conference*, Washington, pp. 650-659, October 1994.

26. B.R. Veillette, *A study of delta-sigma oscillator circuits*, Master Thesis, McGill University, Montréal, August 1995.

27. P. Oleary and F. Maloberti, "A bit stream adder using oversampling," *Electronic Letters*, September 27, 1990, pp. 1708-1709.

28. B.R. Veillette and G.W. Roberts, "FM signal generation using delta-sigma oscillators," *Proc. 1996 IEEE International Symposium on Circuits and Systems*, pp. 1-4, Atlanta, Georgia, May 1996.

29. E.M. Hawrysh and G.W. Roberts, "An integration of memory-based analog signal generation into current DFT architectures," *Proc. 1996 IEEE International Test Conference*, pp. 528-537, October 1996.

30. B. Dufort and G.W. Roberts, "Signal generation using periodic single and multi bit sigma-delta modulated streams," *Proc. 1997 IEEE International Test Conference*, pp. 396-405, Washington DC, November 1997.

31. M. Mahoney, *DSP-Based Testing of Analog and Mixed-Signal Circuits*, IEEE Press, New York, 1987.

32. "IEEE Standard 1057," *IEEE Trial-Use Standard for Digitizing Waveform Recorders*, New York, IEEE, July 1989. (Issued for trial use.)

33. M.F. Toner and G.W. Roberts, "A BIST scheme for an SNR, gain tracking, and frequency response test of a sigma-delta ADC," *IEEE Trans. on Circuits and Systems-II*, vol. 41, no. 12, pp. 1-15, January 1995.

34. M.F. Toner and G.W. Roberts, "A BIST technique for a frequency response and intermodulation distortion test of a sigma-delta ADC," *Proc. 1994 IEEE VLSI Test Symposium*, Cherry Hill, pp. 60 - 65, April, 1994.

35. M. Padmanabhan and K. Martin, "Filter banks for time-recursive implementations of transforms," *IEEE Trans. on Circuits and Systems*, vol. 40, no. 1, pp. 41-50, January 1993.

36. B.R. Veillette and G.W. Roberts, "A built-in self-test strategy for wireless communication systems",

Proc. 1995 IEEE International Test Conference, pp. 930-939, October 1995.

37.   M.F Toner, *MADBIST: A Scheme for Built-In Self-Test of Mixed-Signal Analog-Digital Integrated Circuits*, Ph.D. thesis, November 1995.

Implementing the 1149.4 Standard Mixed-Signal Test Bus

Steve Sunter

T he IEEE 1149.1 Boundary Scan (also known as "JTAG") standard, which defines digital testability structures for implementation on integrated circuits (ICs), has proven to be very successful. It facilitates automated generation of tests for verifying digital IC pin drivers, receivers, and board-level printed wires interconnecting these IC pins. Dozens of companies presently supply automation to implement boundary scan on application-specific ICs (ASICs) and to automate interconnect test for boards containing digital ICs with this standard. As well, the interface is beginning to be used for non-test functions such as in-system programming of field programmable gate arrays (FPGAs).

The 1149.1 standard [20] does not, however, address analog aspects of board test, such as measuring passive component values, detecting opens and shorts between off-chip analog components, and diagnostically testing differential signal paths. The evolving P1149.4 Standard for a Mixed-Signal Test Bus targets these areas, and also provides access for external test of on-chip mixed-signal circuits and for on-chip circuitry to test the rest of an IC or even other ICs, i.e., built-in self-test (BIST). (The "P" in P1149.4 indicates 'Proposed', because it is a proposed standard. After the standard is balloted and accepted, the prefix will be dropped. Henceforth in this text, 1149.4 will be used for conciseness and for clarity when reading this text after the standard has been accepted.)

The two most important features of the 1149.4 mixed-signal test bus are: It is a standard bus, and it conveys continuous variables. Acceptance of 1149.4 by industry will lead to many companies developing automated mixed-signal test solutions based on the standard, which will allow faster time to market for IC and board manufacturers. This will also lead to structured, mixed-signal design for test (DFT) tech-niques oriented around the standard, which can facilitate more-focused development of analog circuit synthesis, as happened for digital circuits. The lack of any quanti-zation of analog signals by the test bus allows the user of the IC to determine the appropriate trade-off between precision and test time, and allows for continuous improvement in this trade-off.

This chapter will first describe the functions, capabilities, costs, and benefits of the 1149.4 standard. Then it will illustrate how to design the standard into an IC effectively, and how to use its capabilities for IC and board-level test. A detailed dis-cussion of sources of measurement error is provided, which will hopefully lead to improved use of the 1149.4 analog bus (rather than discourage its use!). Lastly, the lessons learned from the first two 1149.4 test chips are summarized.

This text is not a substitute for the IEEE 1149.4 document [13], but rather an introduction to allow potential users to assess their need for the test bus, and a prac-tical guide to its implementation. At the time of this writing, the standard had not yet finished the balloting process, so this chapter will focus on the most stable aspects, and on the theory and reasoning in its development. Although the author is presently vice chair of the P1149.4 Working Group, the views expressed here are not neces-sarily those of the Working Group or the IEEE.

7.1 Overview of 1149.1 and 1149.4

Attempts to define a systematic DFT approach for mixed-signal ICs can be traced back to at least 1988 [1],[2], and "1149.4" was reserved as a potential stan-dard around that time. In 1991 the IEEE P1149.4 Working Group officially began developing a standard hardware infrastructure which would address the testing of mixed-signal ICs. To be practical, it needed to be fully compatible with the existing 1149.1 digital test standard. The Working Group soon concluded [3] that an 1149.4 IC needed to be compatible with the 1149.1 standard, and hence the 1149.1 four-pin test access port (TAP) would be required to control all testing and to access the results.

The salient hardware features required for an IC to conform to 1149.1 are illustrated in Figure 7-1 and include: a test access port (TAP) comprising a minimum of four pins (TDI, TDO, TCK, TMS, and optionally, TRST*); a TAP controller; an instruction register and decoder; a boundary data register which links digital bound-ary cells; and any number of optional data registers. Test functions are performed by first sending in bits via the TMS (test mode select) pin, clocked by TCK (test clock), to indicate that an instruction is about to be loaded. The instruction bit sequence is then loaded serially into the instruction register (IR) via the TDI (test data in) pin, clocked by TCK. Each standard instruction then enables various chip-global mode lines, and selects which data register (DR) that test data will next be shifted into or out of. If an instruction selects, for example, the boundary DR, then after an appropriate TMS bit sequence, data can be shifted in serially via the TDI pin and shifted out via TDO (test data out). Each digital function pin has up to 3 bits of the boundary DR, and interprets these bits to cause the pin to perform various test functions such as driving logic 0, 1 or high impedance, and capturing the pin's logic value.

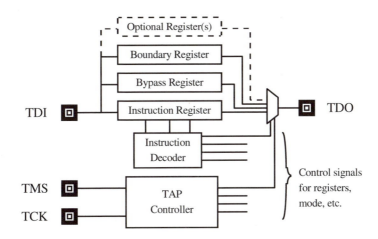

Figure 7-1   *Test Register Structure of 1149.1 IC.*

The hardware features added by 1149.4 (and other features of 1149.1) are illustrated in Figure 7-2 and include: an analog TAP (ATAP) comprising a minimum of two pins (AT1, AT2); on-chip analog bus wires (AB1, AB2); a test bus interface circuit connecting the on-chip and off-chip analog buses; and an analog boundary module (ABM) at each analog pin, included in the boundary DR. An additional chip-global mode line facilitates the one new instruction (*PROBE*), which provides continuous time access to signals while the IC is in mission mode. Off-chip passive or active components are measured by conveying a current (or voltage) from AT1 to any DR-addressed analog pin, and conveying the resulting voltage (or current) from any addressed analog pin to AT2. Generally, the AT1 pins of all ICs on a board are connected to a single bus wire, as are the AT2 pins.

Returning briefly to 1149.1, the advantage of using digital signals is that they can be buffered and sampled any number of times without loss of information (if we ignore timing information, which is an analog parameter). This principle allows 1149.1 boundary scan to effectively test digital circuits, i.e., to observe and control the data at any chosen circuit node (device pin or internal node) via a single input pin ("TDI") and a single output pin ("TDO"). A scan path is used to shift perhaps thousands of bits of data into an IC, apply the data in parallel to selected circuit

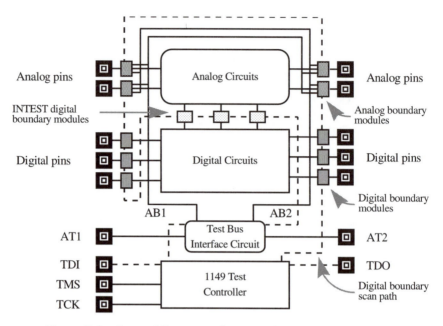

Figure 7-2 *General Structure of 1149.4 IC.*

nodes, sample the thousands of output signals at one instant in parallel, and then serially shift these output bits to a tester (or test circuitry) for comparison with correct logic values.

Analog serial shift registers have been implemented for many years, starting with bucket brigade devices (BBD), then charge-coupled devices (CCD), and lastly current copiers. The circuits usually require a special process (for CCD), or are limited to fewer than 20 shift stages and require many highly linear, low-noise amplifiers. A shift register also requires sampling, which introduces aliasing in the frequency domain.

The two most popular ways to convey a large number of analog test signals in or out through a single pin are a multiplexer and a bus. A bus is, in essence, a distributed multiplexer, but its ultimate analog performance differs. Early proposals for an analog test standard [1],[2],[5] showed liberal use of multiplexers connected in parallel as well as serially. The P1149.4 Working Group chose to require a single wire bus for delivering a stimulus signal ("AT1") to selected circuit nodes, and a single wire bus for routing selected analog output signals to an output pin ("AT2").

Digital boundary scan only needs to deliver either of two discrete voltages and to observe the buffered output logic levels, which are also discrete voltages. However, analog "boundary scan" also needs to measure the value of passive components, in which no buffering occurs. For this reason, the stimulus is typically a current that travels "through" a two-terminal passive component, and the response is the voltage across the component (typically two voltages are measured, and one subtracted from the other).

A second requirement of analog boundary scan is the ability to test open and short circuits in simple interconnect (i.e., printed wires). Studying this requirement revealed that it was already addressed by 1149.1, if the logic levels were re-defined to be the minimum and maximum voltages that analog drivers normally deliver and that analog inputs normally receive.

In summary, 1149.1 provides a standard for testing digital ICs in a system by controlling and observing the pins and internal nodes via a serial, scan path, shift register. The 1149.4 standard extends test capabilities to mixed-signal ICs, by adding two analog bus "wires" which can be used to stimulate and observe analog nodes in the continuous current and voltage domains.

7.2   Test Functions Needed to Implement 1149.4

Conformance to 1149.4 first requires conformance to 1149.1. This allows re-use of existing Boundary Scan Description Language (BSDL) based, test generation software. An analog extension to BSDL will nevertheless be necessary to completely describe all capabilities, but will be developed after the hardware standard passes balloting. This chapter will focus on the hardware aspects of 1149.4 and will not discuss BSDL and other software aspects in any detail.

The mandatory instructions and related hardware functions, common to 1149.1 and 1149.4, are:

BYPASS

- test data bypasses the boundary register, without altering its contents

SAMPLE

- capture a 1-bit binary representation of the voltage at all pins, without affecting the state of the pins

PRELOAD

- shift data through the boundary register, without affecting the state of the pins

PROBE

- selectably convey AT1 current to any analog pin
- selectably convey the voltage at any analog pin to AT2

This is the only 1149.4-specific instruction, and is analogous to the *SAMPLE* instruction. It allows observation of analog signals while the circuit is in normal operation (mission mode). In effect, with appropriate software to program the boundary register, a "virtual guided probe" could be implemented to observe analog waveforms at selected points on the IC's boundary or in its core. Of course, if simple transmission gates are used in the IC, the bandwidth of this "probe" may be far less than a typical bench-top oscilloscope. This instruction also allows injection of current via AT1, though often the normal signal impedances at function pins will mean that only a portion of the stimulus is detected. When CMOS transmission gates are used, AT1 can be used to monitor voltages, hence allowing differential observation in this mode ($V_{AT1} - V_{AT2}$).

EXTEST

- isolate the core circuitry of the IC from the boundary circuitry ("core disconnect")
- selectably drive a logic 0, 1, or high impedance, at any output pin
- capture a 1-bit binary representation of the voltage at all pins, without affecting the state of the pins
- selectably convey AT1 current to any analog pin
- selectably convey the voltage at any analog pin to AT2
- selectably drive a reference voltage at any analog pin

"Logic 1" and "logic 0" for an analog pin are interpreted, in 1149.4, to be the maximum and minimum voltages that the pin normally delivers or receives, and are denoted V_H and V_L, respectively. For a typical CMOS op-amp output, these voltages would be within 0.5 volts of V_{DD} and V_{SS}. For a radio frequency amplifier output, these voltages may be only ±1 volt relative to midway between V_{DD} and V_{SS}. In other words, the normal driver for an analog output pin may be used to deliver V_H and V_L - driving a relatively high impedance logic "high" or "low" into the driver's *input* will generate V_H or V_L at the output. In fact, this is the recommended approach in 1149.4 and 1149.1. Although the functional driver is the best source of V_H and V_L when the load impedance is low (e.g., 50Ω), when the impedance is high (e.g., 50 kΩ) a parallel digital driver should be used which simply drives to V_{DD} or V_{SS}.

The reference voltage, V_G, may be one of V_H or V_L, providing they are of reference quality (see Section 7.7).

The word "selectably" is used to mean a path is enabled when the boundary register is appropriately programmed and after the appropriate falling edge of

TCK. The word "convey" is used because a path could be through CMOS transmission gates or tri-state analog buffers, or a combination of both.

Implementing the high impedance state at any input is typically very simple in CMOS technology – most inputs are already high impedance. However, this may not "isolate" the core circuitry from the boundary. In this context, isolate means that any signal on the function pin must not affect any measured value. If a floating input, or AC signal, were to cause large current swings which affected the power rails and hence caused measurement noise, then isolation would not have been achieved. In such cases, explicitly disabling the input path may be necessary.

The power-down control pin for amplifiers can often be used as a core-disconnect switch for outputs (assuming it makes outputs go tri-state).

The meaning of "high impedance" is interpreted differently in 1149.4 than in 1149.1. Sometimes a low impedance termination (e.g., 50 Ω) is connected on-chip between a function pin and a power rail. It is not usually practical to provide a transistor to disconnect this low impedance – the transistor would be very large and would jeopardize the accuracy of the termination impedance. There is little practical difference between this impedance being on-chip or off-chip, as long as its presence is known. For this reason, such passive impedances are permitted to remain connected to a pin in the high impedance state, as long as it is documented for users of the IC. However, all on-chip AC and DC sources must be disconnected in this state. On-chip sources usually means transistors. Off-chip sources are device pins, such as V_{DD} and V_{SS}.

The optional 1149.4 instructions and related functions for every analog function pin are:

INTEST

- selectably convey AT1 current to any analog pin
- selectably convey the voltage at any analog pin to AT2
- at all inputs to the digital core, deliver boundary register test data, and at outputs, capture the results (including digital signals between the digital and analog circuitry).

IDCODE

- the contents of the IDCODE register are shifted out, indicating IC manufacturer

USERCODE

- the contents of the USERCODE register are shifted out, for user-programma-
  ble ICs

RUNBIST

- isolate the core circuitry of the IC from the pins (i.e. "core disconnect")
- initiate BIST and capture the result in the IR-selected data register

CLAMP

- the present boundary register-driven state of all digital output pins is maintained
- test data bypasses the boundary register, without altering its contents

HIGHZ

- all output pins go to the high impedance state (see special definition under
  EXTEST)
- test data bypasses the boundary register, without altering its contents

One capability that is common to implementing all instructions is the requirement that each new instruction takes effect immediately, without requiring a boundary register update or reset. This means that global Mode signals must enable each
boundary register's ability to control the state of the pins. The Mode signals must
disable the boundary register's control of pins during Test Logic Reset, *BYPASS*,
SAMPLE, *PRELOAD*, *IDCODE*, and *USERCODE*.

Optional analog capabilities, which are not specific to any instruction, include:

- selectably convey the pin's voltage to AT1 (instead of current)
- selectably convey AT2 current to the pin (instead of voltage)
- selectably compare the voltage at the pin to the voltage on AT2
- differentially perform any of the capabilities listed here, also using AT1N and
  AT2N pins

The optional AT1N and AT2N pins are used to access the inverting pin of each
differential pair; AT1 and AT2 are used to access the non-inverting pin of each
differential pair.

Figure 7-3 shows a symbolic schematic of the on-chip circuitry associated
with each analog function pin, the TAP connections, and the analog bus connections.
It must be stressed that this is only a *symbolic* schematic – the switches shown may
not exist in a practical implementation. For example, tri-state analog buffers could

be used instead of switches. The mission-mode function that drives an output pin can be tri-statable; connecting a CMOS transmission gate in series with a low impedance driver is typically impractical – the switch would be larger than the driver, and would degrade performance considerably. Practical implementations for analog output and input pins are shown later in Figures 7-14 and 7-15.

Test Bus Interface Circuit

The interface between the on-chip analog buses and board-level (or multi-chip module-level) analog buses is called the Test Bus Interface Circuit (TBIC); it can be seen in Figure 7-3 and in greater detail later in Figure 7-16. The interface has several important purposes: isolation, calibration, buffering, and 1149.1 compliance. We will describe the basic structure of the TBIC, and show how it facilitates its purposes.

AT1 is a pin of the IC, and AB1 is the on-chip bus. A switch in the TBIC selectably connects the on-chip and off-chip buses. A similar arrangement is provided for AB2 and AT2. In addition, it is possible to access AT1 directly from the AB2 bus, and AT2 directly from the AB1 bus. The latter two switches are essential for characterization. Lastly, switches are provided to connect either on-chip bus to a suitable DC voltage when the bus is not in use to prevent a floating bus from causing interference.

The "switch" between AB2 and AT2 may be a voltage buffer. As will be discussed in considerable detail in Section 7.3 and in Section 7.7, a buffer can facilitate greater measurement accuracy, less impact on the circuit under test, and larger bandwidth. The switch between AT1 and AB1 may be a current buffer. This is especially likely in bipolar technology where transmission gates are not economically feasible. When buffers are connected to either AT1 or AT2, then it becomes clear that to provide a path from AT1 to AT2, two different routes must be provided on-chip: one for current and another for voltage.

Accurate analog measurements are usually based on the difference between the measured value for some reference circuit, which might be as simple as a short circuit, and the measured value for the circuit under test (CUT). If a reference path was provided through a normal function pin, the true reference circuit might be unavoidably connected in parallel with some off-chip low-impedance circuit. For this reason, the reference path is between AT1 and AT2, since these two pins will only have test-related circuitry connected to them whose value is precisely known or which can be measured independent of the mission circuitry. A reference resistor may be connected via relay to AT1 or AT2, within a tester, and its value measured in different ways to determine the parasitic element values of the buses.

The TBIC switches allow the on-chip buses of any ICs that are not involved in a specific measurement to be disconnected, to minimize leakage current, to minimize bus capacitance, and to minimize interference with or by other ICs. This isola-

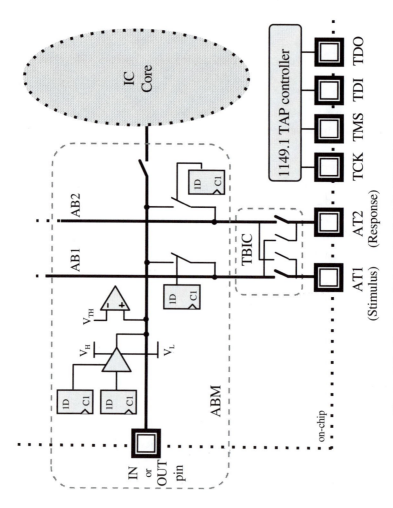

Figure 7-3 *Symbolic Schematic of 1149.4 IC. (Figure courtesy of the IEE, © 1996 [12].)*

tion also allows on-chip self-test via AB1 and AB2, without interfering with other ICs connected to the off-chip buses, AT1 and AT2.

Lastly, switches (not shown in Figure 7-3) to V_{SS} and V_{DD} together with V_{TH} comparators allow the AT1 and AT2 buses to be tested for shorts and opens using conventional 1149.1 boundary scan. This quickly verifies the basic integrity of the two buses, and diagnoses any shorts to any other pins.

It should be noted that when no IC is driving the AT2 or AT1 bus, it will float at an indeterminate level. To prevent this causing problems, a circuit for driving it to a "safe" level should be enabled during mission mode. This is not addressed by the standard because it is not a test-mode function, only one driver should be enabled for a board and only during mission mode, the bus must be driven to a safe voltage, and the driver should be enabled by system-level reset.

Differential Access   An optional extension to the TBIC, which is permitted for differential access, is the implementation of AT1N and AT2N pins. For example, to gain true differential access to a pair of differential analog output pins, the non-inverting output would be connected via AB2 to AT2, and the inverting output would be connected via AB2N to AT2N, as illustrated later in Figure 7-7. Providing the two paths are well matched in their layout on the IC and board, common mode noise can be canceled by the test equipment. Control of the TBIC switches for differential access is similar to that for single-ended access: access to the internal buses is simultaneous, but AT1N and AT2N are controlled independently via a separate TBIC.

Multiple AB1 and AB2 Buses   For large, mixed-signal ICs, there are at least five important reasons why it may not be practical to connect all analog pins and internal nodes of interest to a single pair of AB1 and AB2 buses. We'll first discuss these reasons, and then describe the practical solution permitted by 1149.4.

> a. Too many transistors connected to a single bus increase its capacitance; this is the traditional reason for limiting the number of bus inputs. However, relative to off-chip capacitance, this is not a major consideration for 1149.4.
>
> b. For deep sub-micron technologies (<0.35 µm), the leakage per transistor starts to become significant relative to the stimulus current used in 1149.4. One hundred of these transistors connected to a single bus can cause enough leakage current to affect any 1149.4 measurement. Calibration routines prior to component measurement can account for this leakage to a large degree, but there can also be a significant noise component.
>
> c. For various reasons, sub-micron ICs often have more than one V_{DD} voltage. For example, to facilitate use of very short, CMOS transistor

gate lengths, one V_{DD} may be 3 volts, whereas the interface to off-chip voltages may be 5 volts. In this case, connecting signals to a single AB2 bus may not be possible without risking over-voltage damage to some transistors.

d. An analog bus which is connected to both high frequency and high sensitivity circuitry of an IC may result in unacceptable interference, either due to simple capacitive coupling to the bus wire, or due to capacitive coupling through a disabled CMOS transmission gate.

e. An issue, which will become more important as core-based IC design becomes the norm, arises when several mixed-signal IC designs (or "cores"), containing their own 1149.4 buses, are combined into a single IC design. If each design contains its own TBIC, and the overall design also contains a TBIC, then several transmission gates or amplifiers will be connected in series and the increase in resistance, leakage, and offset may be intolerable. A hierarchy of buses is needed.

The solution to the potential problems described above is the scheme illustrated in Figure 7-4. Many AB2 buses may be connected to a single AT2 pin, each via a different switch (and/or amplifier). All but one AB2 bus will typically be grounded (connected to a DC voltage), and each AB2 bus is separated from other AB2 buses by precisely two switches. In this way, each AB2 (and AB1) bus is more completely isolated from other AB2 (and AB1) buses. Careful observation reveals that this is exactly the scheme used when connecting many ICs on a board or multichip module (MCM).

To access a signal via a multi-AB1 IC, it is necessary to enable the appropriate TBIC switch, in addition to the specific ABM or core switch. The set of ABMs accessible via each TBIC must be documented for users.

In summary, current to or from any node on any IC will travel through only two switches to reach the measurement/test instrument: one at the node itself, and one in a TBIC. Thus, the performance specifications of the analog bus remain in effect, even when many hundreds of nodes are accessed in many cores, ICs, and MCMs.

7.3 Test Capabilities That This Standard Facilitates

The functions described in the previous section can be used to accomplish a wide range of tests, where "tests" include measurement, pass/fail testing, diagnosis, and simple access. Many of the capabilities of a mixed-signal tester are made available at every analog pin. This section will describe the most frequently needed capabilities.

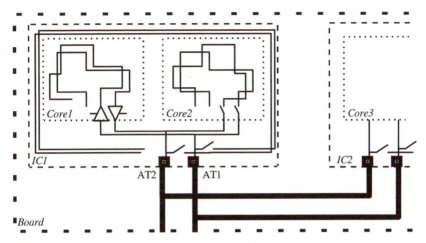

Figure 7-4 *Example of Hierarchical TBIC Connections, Interfacing Three Bus-Pairs.*

7.3.1 Resistance, Capacitance, and Inductance Measurement

Figure 7-5 shows two ICs, each including 1149.4 facilities (not all shown), and an RC circuit off-chip connecting the two ICs. To simplify the following examples, we will assume CMOS ICs with power rails V_{DD} and V_{SS}, and with switches implemented using CMOS transmission gates.

To measure the value of C and R, the following procedure may be followed (referring to Figure 7-5):

  a. Via the boundary scan chain and TAP, enable switches on IC1 so that the AT1 stimulus is routed into the TBIC of IC1 to its AB1 then to pin 1, through R to the capacitor C, and to pin 2 of IC2. Also enable switches on IC1 such that the voltage on pin 1 is connected to AB2 and to AT2, and to an off-chip AC voltmeter. The AT1 stimulus is a 100 μA (peak amplitude) current source in parallel with a 1 kΩ resistance. A simpler alternative is to use a 100 mV voltage source with 1 kΩ output impedance

  b. Measure the amplitude of the voltage at AT1, and divide by the output impedance of the stimulus to determine the stimulus current delivered to AT1. $I_{AT1} = I_{Source-}(V_{AT1}/R_{Source})$, or $I_{AT1} = (V_{AT1}-V_{Source})/R_{Source}$. Specifically, in this example, 100 μA–(96.87 mV/1 kΩ)=3.13 μA AC (peak). Assumed switch impedances are the same as in Figure 7-17.

  c. Measure the amplitude, V_{Pin1}, of the sinusoidal voltage at AT2. The resistance $R_{total} = (V_{Pin1})/(I_{AT1})$, and is equal to the magnitude of the complex impedance between pin 1 and V_{SS}. Specifically, in this example, $(83.2 \text{ mV}/3.13 \text{ μA})=18.8 \text{ kΩ}$. Next, determine the impedance of the capacitor.

d. Via the boundary scan chain and TAP, disable the AT2 and AB2 switches on IC1, and then enable them on IC2, such that the voltage on pin 2 can be observed on the AT2 pin of IC2.

e. Measure the amplitude, V_{Pin2}, of the sinusoidal voltage at AT2. The resistance $R_2 = (V_{Pin2})/(I_{AT1})$, and is equal to the magnitude of the imaginary impedance between pin 2 and V_{SS}, namely the impedance of the capacitor at f = 10 kHz. $C = 1/2\pi f R_2$ Specifically, in this example, $R_2=(70.3 \text{ mV}/3.13 \text{ μA})=15.8 \text{ kΩ}$; $C=1/(2\pi \times 10 \text{ kHz} \times 15.8 \text{ kΩ})=1 \text{ nF}$.

f. Calculate the (Real) resistance, $R^2 = (R_{total}^2 - R_2^2)$. Specifically, in this example, $R^2 = 18800^2 - 15800^2$; $R=10 \text{ kΩ}$

If the capacitance is zero, AC stimulus is still best, to ensure that any constant leakage currents or offset voltages are canceled. This is especially necessary if on-chip voltage buffers are used in the AB2 to AT2 path, or if on-chip current buffers are used in the AT1 to AB1 path. This might be the case when very high bandwidth access is needed, or if bipolar technology is used.

A similar measurement method can be used for measuring inductances. In general, any two-port network of up to three elements can be accurately measured in, at most, two voltage measurements per element. For two-port networks with more than three elements, a more-advanced but well-understood method has been documented [7].

7.3.2   Measuring DC Parameters of Inputs and Outputs

To measure the IOL (or IOH) for a digital output driver, a stimulus current with an AC component is applied to AT1 while the output pin driver of interest is set to drive a logic 0 (or logic 1). For example, an AC current of ±50 μA superimposed on a DC current of 50 μA can be applied, and the voltage amplitude measured will indicate the impedance of the pin driver: Rout = V/(50 μA). The output current for a specific output voltage, e.g., VOL = 0.4 V, can be estimated by extrapolation (for deep sub-micron devices this extrapolation will not be linear).

For analog output drivers, the output impedance while driving midway between V_{DD} and V_{SS} can be similarly determined.

To measure VIL (or VIH) for a digital input, the optional voltage capability of the AB1 bus is needed. A DC voltage can be applied to many inputs simultaneously only if no load is present at any input pin and if the power drawn by the input receivers is not excessive. While applying a specified voltage, e.g., VILmax, the output of every pin's receiver is sampled by the boundary scan register, and compared with the correct value.

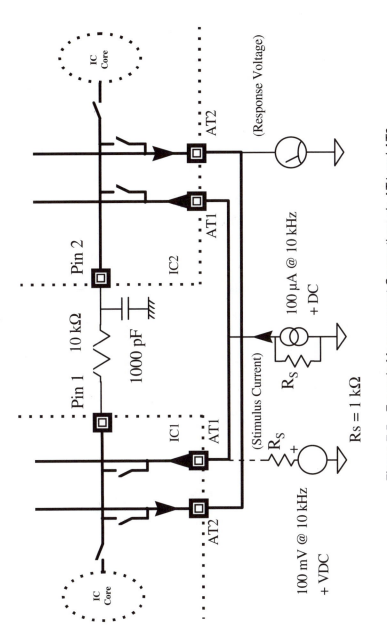

Figure 7-5 Example Measurement Connections via AT1 and AT2.

199

The value of on-chip pull-up and pull-down resistances can obviously be measured via the analog bus just as described for off-chip impedances. Pin input leakages and input impedances can be measured similarly.

Short (and open) circuits between pins, as typically caused by solder (or lack thereof) during board manufacture, are usually first detected using the digital capabilities of the *EXTEST* instruction to apply tests in parallel. Measuring the parametric impact of manufacturing defects, via AT1 and AT2, is performed serially and is therefore much slower.

7.3.3   Differential Measurements

As power supply voltages decrease with each new CMOS process technology, the need for differential analog functions increases, because thermal noise is largely independent of power supply voltage. Differential methods allow 6 dB extra signal swing and almost complete cancellation of most systematic noise sources such as offset, clock noise, and other interference. The ratio of the signal to random (e.g. thermal) noise increases by 3 dB, since noise increases by only 3 dB when two signals with equal noise levels are summed.

The 1149.4 standard allows three ways to access signals differentially.

a. As is possible in 1149.1, the single-ended output of a differential receiver can be sampled as a one-bit digital value by one of the ABM's register stages. This is recommended both for inputs and outputs, using a dedicated test-mode comparator in the latter case, as shown in Figures 7-6 and 7-7.

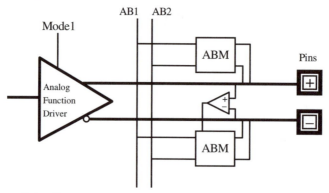

Figure 7-6   *Example of connections to a differential output pin pair.*

b. Use the optional two extra analog bus pins, AT1N and AT2N, as previously mentioned in Section 7.2. Differential signal currents may be applied to a differential pin pair using AT1 and AT1N, and the resultant differential voltage may be monitored via AT2 and AT2N, as shown in Figure 7-7. This facility is likely economic only when most measurements must be done differentially due to a high noise environment, or when most pins of the IC are differential pairs.

c. Use the optional voltage output capability of the AB1 bus, or the current input capability of the AB2 bus. Both capabilities are inherently possible when CMOS transmission gates are used as the on-chip switches. A differential pin pair is observed by enabling the AB2 switch for the positive output pin, and the AB1 switch for the negative output pin, and observing the differential voltage on AT1 and AT2. Similarly, a differential current can be applied. Note, however, that this method does not permit measurement of passive component values because differential voltage cannot be observed at the same time as differential current is applied (on the same bus).

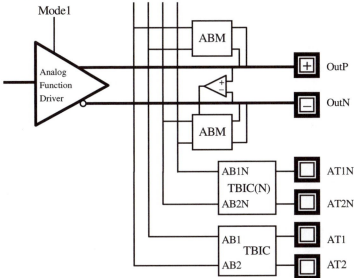

Figure 7-7   *Fully Differential Access to a Differential Pin Pair.*

7.3.4   Bandwidth

An important performance parameter for any analog signal path is its bandwidth. What bandwidth can be expected for the 1149.4 bus for any (future) off-the-shelf IC?

The bandwidth is determined by the following elements (shown in Figure 7-8):

- impedance of the signal source, Z_{CUT}, typically $10\ \Omega - 100\ k\Omega$
- impedance of the switch, R_{ABn}, with a typical value of $2\ k\Omega$, for a CMOS transmission gate
- capacitance of the on-chip bus, C_{ABn}, with a typical value of 1–3 pF
- impedance of the AB2-to-AT2 bus switch, R_{TBIC}, with a typical value of 200 Ω, for a large CMOS transmission gate
- capacitance of the board-level bus, C_{ATn}, including the pin capacitance of all ICs connected to the bus; assume 4 other ICs (5 in all), @ 5 pF each, and 5 pF for the board: the sum is 30 pF
- the input capacitance of the measuring instrument, typically 10 pF

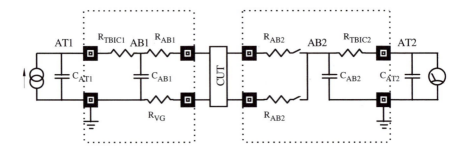

Figure 7-8   *Circuit Elements that Affect Analog Bus Bandwidth.*

Approximating this scenario (when $Z_{CUT} = 0$) as 2200 Ω driving 40 pF, the – 3 dB bandwidth is $1/2\pi\,RC = 1.8$ MHz. For a board-level bus with 10 ICs connected to the AT2 bus, the bandwidth would be 900 kHz. For any IC that uses the 10 $k\Omega$ maximum switch impedance permitted by 1149.4, the bandwidth reduces to less than 200 kHz. Clearly this bandwidth is not sufficient for cellular radio ICs. It is, however, sufficient for audio and many other applications. More importantly, it is sufficient for measurement of passive components when appropriate calibration techniques are used (as described later). Many impedance measuring instruments, with four or more digits of resolution, use an AC stimulus with a frequency less than 10 kHz, which is well within the bandwidth of a future typical 1149.4 IC using simple, low cost, CMOS transmission gates.

For the above typical values, with signal source impedances greater than 20 Ω, the signal level will be reduced *at the source* by more than 1% when it is accessed via the analog test bus. The signal level observed *at AT2* will be reduced by 90% for frequencies greater than 20 MHz. In these cases, on-chip buffers will be needed so that typical off-chip capacitance will not reduce the signal level significantly.

7.3.4.1   *Voltage Buffers*

A common approach to reduce the impact of off-chip capacitance on nodes accessed by a bus, is to connect an on-chip bus to an analog amplifier which drives an output pin [14]. This same approach can be used for the 1149.4 bus: AB2 is connected to a buffer amplifier which drives AT2. Then each node accessed only senses a small, on-chip capacitance. A 50 MHz amplifier was included in one of the P1149.4 test chips [11], though not as a buffer for the test bus. The area for the buffer is relatively large in a 1.5 μm CMOS technology, making cost a primary consideration. Also, cross-coupling is likely when an analog bus with large voltage swings or high frequencies traverses an IC. Shielding may be provided by using other layers of metal between the bus and unrelated wires, but this may be too costly in reducing interconnect flexibility. A better solution might be constraining the swing of the analog bus by using low-gain analog buffers, and calibrating the gain before performing any measurements.

One such method has been reported [15], in which digital tri-state drivers are used as 100 MHz linearized voltage-to-current buffers, as shown in Figure 7-9. By

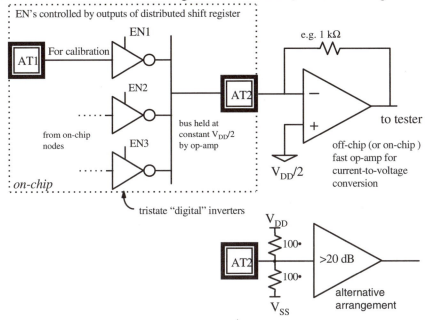

Figure 7-9 Circuit Technique for Area-Efficient High-Frequency Analog Bus. (Figure courtesy of the IEEE, © 1994 [15].)

connecting the bus to an off-chip (or on-chip) virtual ground or 50 Ω load resistance terminated by a voltage midway between V_{DD} and V_{SS}, and amplifying the resulting signal, the signal travels on the bus as a current. The effective output impedance of the buffer becomes 50 Ω (or less, when a virtual ground is used), increasing the bandwidth greatly. For large swings, the buffer has only 5 - 10% non-linearity and it can be mostly canceled by first determining the transfer function of a representative buffer on each IC, and then applying the complementary function (in tester software) to each signal output. For 100 mV signal swings, the non-linearity may be less than 1%, but the gain will depend on the DC bias voltage. If an off-chip load resistor and amplifier is used, then a wide choice of high performance amplifiers is available, and the IC's area is not increased.

7.3.4.2   Sample-and-Hold

A technique for observing repetitive, high frequency, analog waveforms is to use a sample-and-hold circuit to sample the waveform at a rate slower than the frequency of interest. This is known as under-sampling. Its performance is primarily limited by jitter in the sampling clock and by the speed of the sample-and-hold buffer. A 500 MHz effective "bandwidth" has been reported [16]. The strobe signal can be a dedicated digital clock or an edge of the TMS signal (see Strobed Comparators below). The sample-and-hold capacitance can even be the AT2 bus capacitance itself, though an op amp buffer is best. Under-sampling is vulnerable to aliasing in the frequency domain, but if appropriate precautions are taken, its simplicity and performance are impressive.

7.3.4.3   Strobed Comparators

In conventional automatic test equipment (ATE), high-speed digital waveforms are measured using strobed comparators. The signal of interest is connected to the comparator, as is a variable reference voltage. The comparator latches the result of comparing these two signals when a precisely-timed, digital, strobe signal transition occurs. By repeating the stimulus vector sequence, and adjusting the time of the control signal transition, a binary search is performed until the exact transition time at the reference voltage level is determined. In [18], this scheme is exploited within the 1149.4 framework, in a method called "Early Capture" and shown in Figure 7-10. Transitions of the TMS signal, which is one of the 1149.1 TAP signals, may be used to strobe the comparator, since transitions on TMS are ignored by the TAP. The level of the TMS signal is normally only monitored by the TAP during a rising transition of TCK. The voltage at which the delay is measured (i.e., the reference voltage for the comparator), may be routed to the comparator on AB1 or AB2.

This technique allows off-chip observation of repetitive high frequency, analog waveforms: at each strobe time, the voltage of the sensed signal is determined by adjusting the reference voltage on the AB1 or AB2 bus in a binary search fashion, each time reading the comparison result via the boundary scan register; then the strobe time is increased by a constant, known increment, and the voltage is again determined. Eventually, an entire voltage versus time plot can be generated with an effective bandwidth limited only by the jitter in the strobe instant. If a constant reference voltage is used, the Early Capture technique can also be used for simple delay measurements, with sub-nanosecond accuracy.

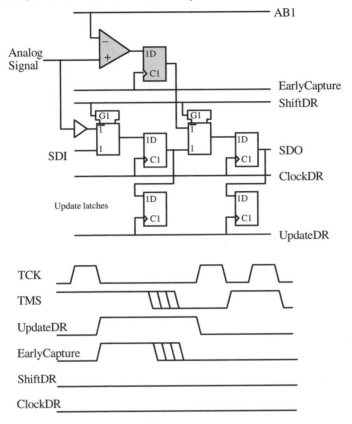

Figure 7-10 *Early capture scheme for measuring high bandwidth repetitive signals.(Only the shaded gates are added: The other gates belong to standard ABM).*

7.3.5   Delay Measurement

Since the analog test bus has no inherent sampling, *in theory* it is possible to simply access signals of interest and deduce the delay between them. To ensure constant delay offset, an off-chip timing edge is used as a reference, then the delay to one on-chip signal is measured (by enabling a switch between the signal of interest and AB2), and then the delay to the second on-chip signal is measured (by enabling an alternate switch). For relatively long delays, one signal could be accessed via AB1 and the other, simultaneously, via AB2.

In practice, however, for typical digital delays, the 2 MHz typical bandwidth of the bus limits the minimum measurable delay to approximately 1 µs. There are at least three ways to circumvent this limit: the simplest is to use digital tri-state buffers, the most general is to increase the bandwidth by using high-speed linear buffers (discussed in previous section), and a reasonable compromise is to use a strobed comparator (discussed in previous section).

Using conventional tri-state digital buffers to drive the analog bus in the IC design is simple. These can be enabled by "private" data register bits in the boundary or in an internal scan register. A digital buffer located in the core of the IC typically has an output impedance on the order of 500 Ω. When driving a total on-chip and off-chip bus capacitance of 35 pF, a rise time of 5–10 ns can be expected, which limits the minimum measurable delay to approximately 10 ns.

7.4 Potential benefits of using this standard

The specific benefits of using this standard mixed-signal test bus are many, given the functions and capabilities described in the previous sections. The primary benefits, which were the objectives of the Working Group developing the standard, are:

1. Reduce the *cost* of verifying board-level interconnect and passive components for dense boards or MCMs containing mixed-signal ICs, by reducing or eliminating the need for bed-of-nails testing.

2. Reduce the test development *time* for mixed-signal boards by allowing test automation software to be developed and re-used. Test automation companies can develop test solutions which are potentially cheaper for any individual IC designer or manufacturer than company-internal test solutions, because the costs are shared by a wider customer base, and because the company focuses only on test automation.

Other benefits, which were realized after the Working Group began, are:

3. True differential access to signal pairs allows accurate testing and diagnosis of low-level signals in a noisy environment.

4. Reduced pin-count IC test, at the wafer, package, and board level, allows the use of lower-cost testers and probe cards. Design companies could afford to supply low-cost personal computer-based testers to designers and test engineers, allowing them to more efficiently develop and debug production tests.

5. Focusing industry's creativity on a single, standard test interface can result in better testability than any *ad hoc* approach, even compared to company-internal standards. An excellent example of this focused creativity was seen in the Internet, where a relatively low bandwidth medium is now used for the trans-

mission of text, digital audio, high-resolution graphics, and video. No single company could have achieved this progress in such a short time.

6. An electronic access to components which are embedded in mechanical structures is more efficient than providing special test points or requiring partial disassembly. The structures are typically needed for heat dissipation, electromagnetic interference (EMI) protection, or simply mechanical strength. Unfortunately they can prevent quick access to signal pins, or their removal corrupts the signal measurement.

7. An on-chip test bus facilitates Built-In Self-Test (BIST), because the test bus allows a central test facility to easily access the rest of an IC or board. BIST can further reduce test development time and test cost, and can be re-used at each higher level of integration, at burn-in, and in the field.

7.5 Costs of Implementing This Standard on an IC

There are several costs that should be considered when implementing this standard (or any circuit) on an IC. Whereas area for additional gates is usually considered first, the dominant cost increase incurred to include 1149.4 is usually the cost of the six additional pins: TCK, TMS, TDI, TDO, AT1, and AT2. First, let us discuss the details of each cost increase [10],[12].

Extra Pins   To add 1149.4 to a typical mixed-signal IC requires the minimum addition of six additional pins. These pins are mandatory and cannot normally be multiplexed with any function pins since that would preclude test access while in function mode. For ICs which already include 1149.1 boundary scan, only two additional pins are needed: AT1 and AT2. If optional pins are implemented, TRST*, AT1N, and AT2N, then the pin count increase rises to nine. To determine the impact on die size, the bonding pad pitch must be known. The cost increment decreases as the pad pitch decreases. Often, the die size is limited by the area of the circuitry in the core of the IC, and in these cases adding more bond pads does not require any increase in die size.

An alternative bond pad arrangement is the bond pad array. The pads are arranged in an array, uniformly-spaced in the X and Y direction, over top of the logic gates. The inverted die is then attached to the die package, using small metal (e.g., solder) bumps, one at each pad location. The cost of additional pins might be less, relative to peripheral bond pad placement, depending on the pad pitch used.

Extra Gates   The number of logic gates required to implement a typical TAP controller is approximately 500 [19]. Approximately 60 additional logic gates are also needed for each pin that is to have an ABM (analog boundary module). These per-pin gates may be located immediately adjacent to the controlled function pin, or may be placed as random logic in the core of the IC. Lastly, two electro-static discharge (ESD) protected paths are needed to each bonding pad to be accessed: one for the stimulus current ("force"), and the other for the voltage monitor ("sense") path.

Clearly the area occupied by the additional logic gates decreases for smaller geometry processes. This was a consideration when the Working Group voted [17] to increase the number of register bits per ABM from three to four – the area in a 0.35 μm CMOS process increased from 0.0082 mm^2 to 0.009 mm^2 per ABM, a 9% increase per ABM but only 0.077% relative to a 1 cm^2 IC.   The area in a 0.8 μm process is, of course, more significant but still small.

Performance Impact   The 1149.4 test standard apparently requires the addition of five switches to each function pin. As mentioned in previous sections, one of these switches may be implemented conceptually, with tri-state buffers, with high impedance mode in an output driver, or for some inputs, with no switch at all. Even if the access switches are all implemented as CMOS transmission gates, the additional capacitance of a disabled switch is usually negligible relative to the capacitance of the bond pad itself. When the switch is enabled, however, the capacitance of the analog bus must be considered, and this can be greater than 20 pF as previously shown. If that capacitance (in series with the switch resistance) is deleterious, then a voltage buffer may be used, as discussed earlier.

Extra Complexity   Adding any circuitry to a well-designed function always increases complexity. If a mixed-signal circuit is designed without anticipating 1149.4, then the following design changes are the most significant that may be needed:

- make all output pin drivers tri-statable
- for output drivers with less than 10 kΩ output impedance, insert a multiplexer at the input to the driver so that a logic 1 or logic 0 can be driven into the driver in test mode, and the normal analog signal in mission mode
- for output drivers with greater than 10 kΩ output impedance, a simple digital buffer can be connected to the output pin and enabled in test mode
- for inputs, ensure that in test mode, any level of input signal will have no effect on any pin measurements; if noise or high I_{DD} might result, then ensure that the receiver is disabled (unless small signal swings are normal at the input, in which case the output of the receiver might need to be monitored to detect logic 0 and logic 1 inputs – hence, disable subsequent circuitry)
- for resistive ESD and latch-up protection circuitry, ensure that two separate protected paths are connected to the bonding pad, one for stimulus current (and output functions) and one for voltage monitoring (and input functions)
- to control the above functions, provide ABM digital circuitry either at each pad site or amongst random logic in the core of the IC

• provide a TAP controller circuit

• provide 6 pins (TCK, TMS, TDI, TDO, AT1, AT2, and possibly others)

The increase in complexity is obviously less if, for example, outputs are originally designed to be tri-statable, or if every bonding pad is designed to have two ESD-protected signal paths.

For high-frequency circuits, care must be taken to route the global test control signals away from sensitive signals, or to provide grounded shielding in between signal nodes and test lines. It may be necessary to make AB1 and AB2 analog switches using T-networks (see Figure 7-11): two transmission gates in series with a common control signal, and a transmission gate connected between analog ground (or V_{SS} or V_{DD}) and the middle node of the two series gates, and whose control signal is the inverse of the common control signal. These switches require more area and design effort, but ensure better isolation for high-frequency signals when they are turned off.

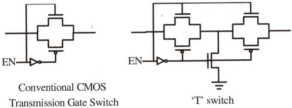

Conventional CMOS
Transmission Gate Switch

'T' switch

Figure 7-11   *T' Switch, Compared to Standard CMOS Transmission Gate.*

Summing the Costs   For a "typical" mixed-signal IC with 68 pins, the cost of adding 6 pins is relatively large, which is not surprising to most designers. What is surprising is that adding 6 pins to a 400 pin IC can increase the cost more in dollars *and* proportionally. Figure 7-12 shows the area and cost increase for an IC initially without any boundary scan, and also for an IC which initially has 1149.1 boundary scan, for a 0.8 µm process. Figure 7-13 shows the predicted increases for a 0.35 µm process and 100 µm pad pitch. Figures 7-12 and 7-13 were carefully calculated using a specific set of assumptions [12], but are only intended to be illustrative – the costs will be different for a technology with different layout assumptions, different defectivity, and different wafer and package costs.

An alternative to embedding 1149.4 in every IC on a board is to use an IC whose only purpose is providing 1149.4 access to board-level signals [9]. Such ICs might be made in higher volumes and more cheaply, per signal accessed, and could be used for accessing pins of ICs which cannot economically justify inclusion of 1149.4. This approach would, however, impose greater wiring complexity at the board-level instead of at the IC-level.

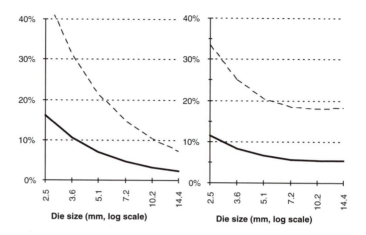

Figure 7-12   *Area increase versus die size, for 0.8 μm technology, 200 μm pad pitch, relative to the IC with (solid line) and without (dashed line) 1149.1. (Figure courtesy of the IEE, © 1996 [12].)*

In conclusion, it is clear from examination of Figures 7-12 and 7-13 that as

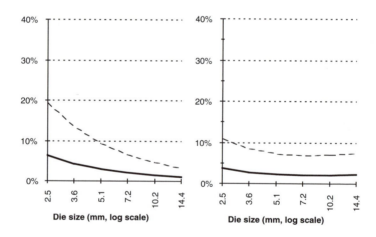

Figure 7-13   *Area increase versus die size, for 0.35 μm technology, 100 μm pad pitch, relative to the IC with (solid line) and without (dashed line) 1149.1. (Figure courtesy of the IEE, © 1996 [12].)*

pad pitch and area per gate decrease, the cost of adding 1149.1 and 1149.4 is decreasing to the level of a few percent. Meanwhile the benefits of greater access to higher-density circuitry are increasing to the point where standard test access simply becomes essential.

7.6   Practical Circuits Compliant with the Standard (Draft 18)

The following three figures contain examples of practical implementations of 1149.4 for analog function pins. In these examples, the analog driver could be a digital driver – there would be no difference in the test circuitry added to be compliant with mandatory rules of 1149.1 and 1149.4. The third example is for the TBIC, including the AT1 and AT2 pins.

Figure 7-14 illustrates a simple implementation for an analog output. The low impedance analog output driver is also used to drive Vmin (logic 0) and Vmax (logic 1) in test mode, by suitably driving its input. A second ESD-protected path has been added for monitoring the voltage via AB2.

Figure 7-15 shows an analog input which has no explicit core-disconnect switch since any input level has no effect on the internal operation of this IC. A simple, digital, tri-state driver suffices for driving logic 1 and 0. The input comparator for simple interconnect test is a multiplexer. A simple D-type with multiplexer input can be used as the test-mode comparator, providing that it does not draw excessive current when enabled while the input signal is midway between V_{DD} and V_{SS} (i.e., don't use a high-drive gate).   The advantage of this approach is elimination of the need for a reference voltage level, since the reference is inherently approximately $V_{DD}/2$. If the comparison is only enabled when ShiftDR and ClockDR are low, then the correct value will be latched as ClockDR goes high (hold time greater than 0), and no power will be consumed in mission mode or other modes. An Exor gate has been included in this implementation to improve digital fault coverage by observing the logic levels driving the AB1 and AB2 switches.

Figure 7-16 illustrates an implementation of the TBIC for an IC with many sensitive signals and thus requiring a voltage buffer for AT2.

7.7   Achieving Measurement Accuracy

In engineering, the commonly accepted threshold for a parametric effect or error being significant, is 10%, i.e., an order of magnitude. For a test to measure these parametric effects, it should be at least another order of magnitude more accurate than 10%. Using this reasoning, a minimally acceptable level of accuracy for a test is 1%. The parametric specifications for the 1149.4 bus were arrived at with the objective of all sources of error, both systematic and random, adding up to less than 1%, when measuring board-level impedances from 10 Ω to 100 kΩ. This degree of accuracy can easily be improved upon in many circumstances, but will not be practical in others. Outside this impedance range, accuracy should degrade predictably and gracefully.

This section will first discuss the key parametric specifications of the analog bus, and will then address all its known sources of DC and AC measurement error

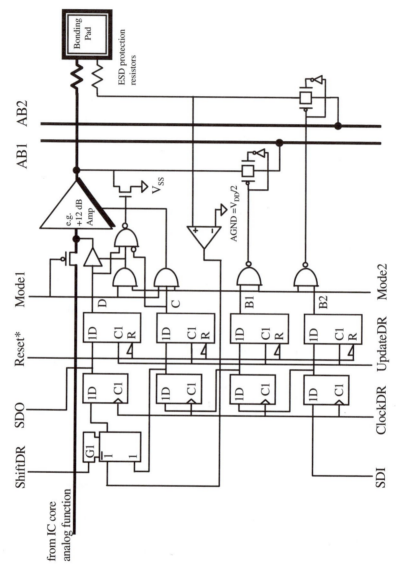

Figure 7-14 *Practical Schematic for Analog Output Pin of 1149.4 IC. (Figure courtesy of the IEE, © 1996 [12].)*

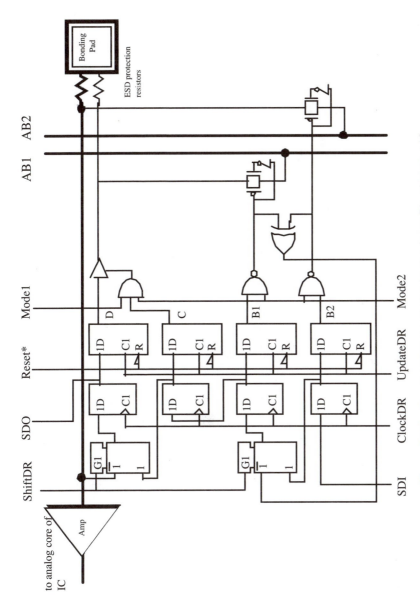

Figure 7-15 Practical Schematic for Analog Input Pin of 1149.4 IC.

213

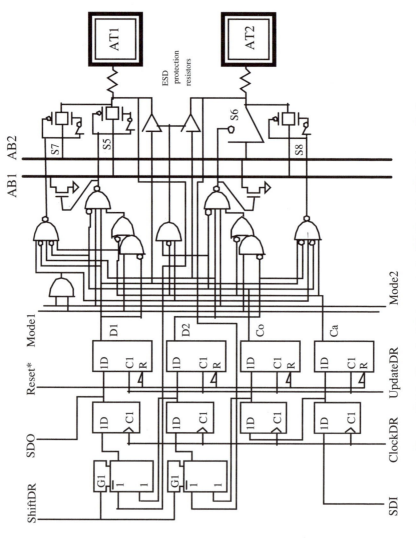

Figure 7-16 *Practical Schematic for TBIC with AT2 Voltage Buffer.*

and how to account for them. Most measurement errors are caused by non-infinite impedances, non-zero impedances, nonlinearity, or noise (in order of significance).

Maximum Switch Impedance   The best estimate of the value of the board-level impedance connected to a specific analog output pin is the inherent impedance of the pin driver itself, for this is the most likely value of the load. For this impedance the impact of measurement error will be half its maximum value. For example, a 1% measurement error in the value of a 600 Ω output load connected to a pin with 600 Ω output impedance causes a 0.5% error in the estimated gain of the circuit. But if the load impedance is 6000 Ω, then a 1% measurement error only causes a 0.05% error in the estimated gain. It is unlikely that this pin would be connected to a 60 Ω load, but if it was, a 1% measurement error will cause a 0.95% error in the estimated gain.

If we assume that a meter requires the maximum permissible stimulus current to flow through the component or circuit under test (CUT) so as to generate the maximum possible voltage, then connecting to the CUT via a switch impedance having a value equal to that of the CUT, will reduce the obtainable accuracy by half. Any further increase in the switch's impedance will increase the measurement error by the ratio of the impedances, i.e.,

R_{SW} / R_{CUT} = 0.5 x measurement_accuracy / meter_accuracy

If we assume that the value of R_{CUT} is R_{PIN}, then

R_{SW} < 0.5 x R_{PIN} x measurement_accuracy / meter_accuracy

Low-cost, test-quality voltmeters with 4 digits of resolution achieve 0.01% accuracy. Therefore, to achieve 1% measurement accuracy,

R_{SW} < 50 x R_{PIN},

where R_{SW} is measured between AT1 and the function pin; R_{SW} includes the AT1–AB1 switch in the TBIC, and the AB1 switch in the function pin's ABM; the CUT is connected between the function pin and ground.

What about the case when the output is that of an operational amplifier which uses feedback to achieve an effective output impedance of approximately 0, but only for load impedances greater than, say, 50 Ω. Input pin and bandwidth considerations were used to formulate a more general 1149.4 rule.

For CMOS input pins, R_{PIN} could easily be greater than 1 MΩ. Using the above guideline leads to an absurdly high maximum value, and so a maximum value for R_{SW} was set at 10 kΩ. Making CMOS switches with greater impedances than this requires more on-chip area anyway, so this is a practical upper limit. Also, bandwidth considerations make 10 kΩ just tolerable.

It is worth recalling at this point, that the switch need not be a CMOS transmission gate, as is often assumed. The switch may be a buffer amplifier, in which case the above discussion about impedance is best ignored! For the AT1 to function-pin path, if current buffers are used, their output impedance is ideally infinite, and in practice, much greater than 10 kΩ. For the function pin to AT2 path, if voltage buffers are used, their output impedance is ideally zero, and in practice, much less than 100 Ω (when not current limited).

Maximum Stimulus Current   The maximum stimulus current that an 1149.4 IC is required to be capable of accurately conveying from AT1 to any function pin is ±100 μA. This value was somewhat arbitrarily chosen because it is practical for today's measurement equipment and for past, present, and (foreseeable) future IC technologies. Mandating a larger stimulus current-handling capability would require greater linearity in the impedance of switches or lower values of switch impedance. Of course, some IC tests might need greater current range (e.g., measuring DC parametrics of high-drive pins via AT1 and AT2); this is entirely permissible within 1149.4.

Maximum Response Voltage   The maximum response voltage that a 1149.4 IC must accurately convey from any function pin to AT2 is ±100 mV plus any DC voltage between V_{DD} and V_{SS}.

This rule ensures that the voltage generated in response to the stimulus current can be measured for a wide range, even when the voltage is across a capacitor or inductor connected to V_G via a switch. In Figure 7-17 the voltage at the node (Pin 2) connecting the V_G switch to the capacitor will vary from V_G–100mV to V_G+100mV when only 50 μA (peak) AC stimulus current is used, regardless of the DC bias applied to the stimulus. Also, in Figure 7-17, we can see why the voltage swing must be constrained to ±100 mV: the current through the pin protection diodes (ubiquitous in CMOS) will increase exponentially with voltage, and can exceed a milliamp when the voltage approaches the value of 0.7 V.

Notice that if 100 μA stimulus current is used, and the total switch impedance is the maximum value of 10 kΩ, then there will be a 1 V drop in potential across the switches. If we assume that the V_G switch impedance is 4 kΩ, the maximum current that can be safely used to stimulate a capacitor connected to V_G is 25 μA. Fortunately, this limitation is not unreasonable for present technologies and test equipment.

To facilitate the use of stimulus currents less than 100 μA, and the often very small response voltages, the stability of the reference voltage V_G must be commensurate. Typically, V_{SS} will be used, in which case stability is almost totally determined by the user of an IC. Noise on V_{SS}, that is not canceled by the measurement equipment, will translate into measurement error.

7.7.1   DC Measurement Errors

Kelvin or 4-Point Probe Measurement   A range of switch impedances is permitted, but it is not necessary that the exact value of the switch resistance be

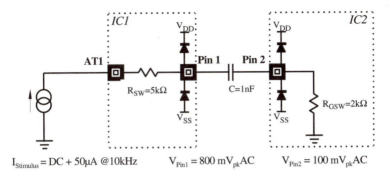

$I_{Stimulus} = DC + 50\mu A$ @10kHz $V_{Pin1} = 800\ mV_{pk}AC$ $V_{Pin2} = 100\ mV_{pk}AC$

Figure 7-17 *Voltages When Measuring a Capacitance between Pins. (Note that a larger AC stimulus would cause pin2 peak AC voltage to exceed 0.7V.)*

known. This is the most important characteristic of the 1149.4 metrology. By using the 4-point probe or Kelvin-style of measurement, the impedance of the switches can almost be ignored for DC measurements. The left side of Figure 7-18 shows the principle as it is used when measuring the resistance per unit length across the surface of a silicon wafer, and the right-hand side shows how it is implemented through an IC via 1149.4 buses.

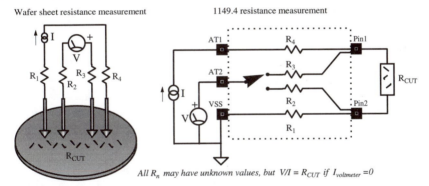

All R_n may have unknown values, but $V/I = R_{CUT}$ if $I_{voltmeter} = 0$

Figure 7-18 *Four-probe Measurement Technique for Measuring Silicon Wafer Resistivity Compared to 1149.4 Measurement.*

As seen in Figure 7-18, a known stimulus current, I, enters the CUT through one "probe" which has unknown series resistance, R4, and exits the CUT through a second probe which also has an unknown series resistance, R1. Two other probes are connected across the CUT, each having unknown series resistance, and the voltage, V, is measured using a high input-resistance voltmeter. Since all of the stimulus current can be assumed to travel through the CUT, and a negligible amount enters the voltmeter, the resistance of the CUT is simply the voltage divided by the current (V/I). We need to ensure that series resistance of the probe resistances plus that of the CUT does not exceed the voltage limit of the current source. This scheme is used by commercial test equipment for performing the most precise measurements, even when the probe impedances are known with some precision. The voltage can be measured very precisely because it is done differentially.

The primary difference between the 4-point probe scheme just described and the 1149.4 scheme is that the two voltages cannot be measured simultaneously. The voltages are measured one at a time, but accuracy may be improved by averaging the noise over a longer period of time (e.g., 1–10 ms).

An important consideration when designing an 1149.4 IC is the correct connection of the stimulus current path and the voltage sensing path. The impedance measured in both circuits of Figure 7-18 is between two junction points. Care must be taken to ensure that the junction in an IC occurs as close to the pin as possible. This is easily achieved by making the bond pad itself the junction point – separate ESD protection resistances should be used for the AB1 and AB2 paths, as shown in Figure 7-19, example C. Examples A and B of Figure 7-19 show examples of shared resistance (A) and incorrectly connected resistances (B).

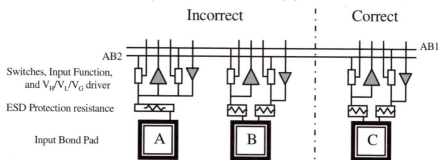

Figure 7-19 *Good and Bad Examples of AB1 and AB2 Connections to the Bond Pad.*

 a. A. single resistive path connects drive and sense path to pin

 b. B. separate paths, but V_G driver and AB2 (sense path) on same resistor

 c. C. correct; all outputs share one resistance, all inputs share the other

Switch Impedance Linearity When typical CMOS transmission gates are used to implement the switches used in the AB1 and AB2 paths, the linearity of the switches is a consideration. The impedance of a CMOS transmission gate is very dependent on the DC bias voltage of the signal being transmitted. Figure 7-20 shows a typical plot of small-signal resistance versus bias voltage (if a large signal is used, then exact wave shape becomes important). For deep sub-micron devices, this non-linearity can be even more pronounced. For this reason, only small currents and small voltage swings should be used, i.e., less than 100 μA and less than 100 mV. If larger swings are used, then the non-linearity will cause the apparent amplitude of the AC signal to decrease or increase, resulting in measurement error. If a DC stimulus is used (with longer settling time and greater sensitivity to offset voltage), then the non-linearity becomes irrelevant.

The 1149.4 standard does not mandate a limit for non-linearity, but it should be less than 0.1% in any 200 mV range, for any DC voltage bias, with a resistive load on AT2 greater than 1 MΩ. This requires the switch impedance to vary by less than 1 kW for any 200mV range, which is easily achievable in practice and obviates the need to characterize linearity before performing measurements.

Canceling DC Offset Voltages To reduce the impact of any 1/f noise (i.e. noise whose amplitude is inversely proportional to frequency) including any DC offset voltages in the test circuitry itself, an AC (alternating current) stimulus can be used – any frequency between 1 kHz and 10 kHz is sufficient. At frequencies lower than 1 kHz, 1/f noise will become significant, and settling times will become excessive. At frequencies higher than 10 kHz, the finite bandwidth of the analog bus may attenuate the signal by more than 0.3%. The voltage should be sensed after first filtering with a suitable band-pass filter, whose bandwidth is at least 10 kHz to ensure a reasonably fast settling time (e.g., a millisecond).

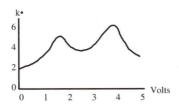

Figure 7-20 *Resistance versus Voltage for Example CMOS Transmission Gate.*

Using AC stimulus, almost any typical voltage offset can be canceled. Of course, if voltage offset becomes excessive (unintentionally, or by design), then a large voltage swing across the CUT may cause clipping. For this reason, the bias voltage or offset should be measured or characterized before performing any measurements of the CUT.

Leakage Currents There are several typical causes of "DC" leakage current on the analog buses; some of them can be canceled, but some cannot. The 1149.4 standard only mandates that anything that prevents 1% accuracy from being achieved, after bus characterization, be documented for the IC user.

Every diffusion connected to the bus (e.g., at every CMOS switch) is actually a back-biased diode and will leak a small amount of current into the substrate. Large diodes are typically used as protection devices at each bonding pad on the IC, and occasionally parasitic bipolar transistors (inherent in most CMOS processes) are also used. In bipolar technologies, leakage is inherent in every transistor node. In all these cases, the leakage is sensitive to temperature and should be re-measured as junction temperatures change during test or in operation.

In addition, depending on the threshold voltage of the transistors in a CMOS switch, a small amount of sub-threshold leakage will occur. The sub-threshold leakage is primarily dependent on the actual value of the threshold. Even though the intended threshold voltage, V_T, for a 5-volt V_{DD}, n-channel transistor may be 0.7 volts, the actual value as a result of normal statistical variations in the process may be as low as 0.5 volts causing a sub-threshold leakage less than a nanoamp. For deep sub-micron MOS transistors (e.g., 0.35 μm), V_T may be as low as 0.3 volts, and with statistical variations might result in a decrease to 0.2 volts. Sub-threshold leakage can be as high as a hundred nanoamps per transistor, and increases with temperature. With perhaps 50 transistors connected to an on-chip bus, the total leakage could be a few microamps, but for 5-volt processes the leakage will typically be less than a nanoamp.

First, the leakage current for AT1 alone can be determined by measuring it when all ICs on the bus have their TBIC switches disabled. Next, the leakage current for AB1 of an IC can be determined by enabling only its TBIC switch, and no other function pins or other IC's TBICs switches. Lastly, a compensating leakage current is added to any current delivered as a stimulus on AT1.

Initial Conditions   As anyone who has performed circuit simulations knows, initial conditions can dramatically alter measured values until a steady-state response has been achieved. This is especially true when measuring board-level capacitances and inductances at production test speeds with kilohm switches and microamp currents.

Whenever capacitances or inductances are measured, time must be given to allow for the initial voltage or current to discharge. The time for a capacitance of 1 μF to discharge 1 volt with a DC stimulus of 10 μA is easily calculated to be 100 ms (t=CV/I), which is the delay necessary before a DC measurement can be performed. However, an AC measurement can be performed almost immediately. The wait time is longer when using DC stimulus than for AC stimulus, because AC stimulus allows measurement in a bandwidth which does not include DC and hence ignores relatively slowly settling DC levels.

If a capacitor is connected between a pin presently driving a voltage equal to V_{SS} and an output pin which is presently driving a voltage equal to V_{DD}, and then the V_{SS} terminal of the capacitor is suddenly driven to mid-rail during measurement of a resistor on that same pin, the other end of the capacitor will be driven above V_{DD}. This will activate the protection diode of the IC which will conduct AC and DC current until the capacitor is discharged below V_{DD}. During this time, measurement of the resistor or capacitor will give incorrect results. If no protection diode is present it is also possible (though not likely, given the small current involved) that voltage damage could occur.

It is worth noting here that several analog bus schemes have been published and proposed as alternatives to 1149.4. The primary reason for them being unacceptable was their sharing use of the analog bus with digital circuitry. This made these schemes incompatible with existing 1149.1 ICs. In some cases, sharing the stimulus bus also meant that every time the bus was used to convey test data to change the voltage monitoring point, the CUT needed to re-initialize due to the interruption in stimulus. The impact on test time would be unacceptable in many cases, since initialization is one of the most important reasons for the length of analog tests relative to digital tests (for circuits of similar complexity).

7.7.2 AC Measurement Errors

The measurement errors that can occur when using a DC stimulus have been discussed. Many of these errors can be easily canceled or reduced by using AC stimulus. However, AC has its own set of measurement errors, which in many respects are more complicated to cancel or reduce.

Capacitive Coupling Between AT1 and AT2 At low frequencies, any capacitance between AT1 and AT2 will prevent the accurate measurement of small, complex, impedances having values similar to the coupling capacitance (or resistance). At frequencies above 10 MHz, the effects can be more dramatic, as reported in [15], because a feed-forward path will be in parallel with the CUT, causing partial cancellation or amplification of the stimulus signal. The diagrams in this chapter show AT1 and AT2 as adjacent pins only to simplify the schematics. AT1 and AT2 should not be adjacent pins on an IC, and should have an AC ground pin (or pins) between them, such as a power supply, ground, or voltage references.

Frequency-Attenuated Access Due to Switch Impedance The bandwidth of the 1149.4 analog bus is determined by the bus's resistances (switches and wires) and by the bus's capacitances (at each switch, the bus wires on-chip and off-chip, the IC's pin, and the voltage measuring unit).

For a bus driven by an IC through switches with R_{SW}=10 kΩ (maximum permitted) and a total bus capacitance of 32 pF, the −3 dB bandwidth will be 500 kHz and the gain at 100 kHz will be 0.980. For another IC on this bus, the switch resistance might be 5 kΩ, giving a bandwidth of 1 MHz and gain at 100 kHz of 0.995. The difference in gain is 1.5%. Since we do not know the bandwidth in advance for an IC driving the bus, we can only assume that the maximum switch impedance is 10 kΩ, and we could therefore have a measurement error of 1.5%.

For a more extreme example, which is still realistic, consider a total bus capacitance of 120 pF (e.g., 10 ICs @ 8 pF per pin, 20 pF for the bus, plus 20 pF for the tester) driven through R_{SW}=1 to 10 kΩ: the –3 dB bandwidth will range from 133 to 1330 kHz, and the gain at 100 kHz will range from 0.798 to 0.997, potentially causing a 20% measurement error.

The simplest solution is performing the measurement with a stimulus frequency less than or equal to 10 kHz to ensure that the bandwidth-related measurement error does not exceed 0.3%. When measuring capacitances or inductances, lower frequencies also increase the generated voltage difference. Another board-level solution is to connect only four or fewer ICs to each AT2 bus wire, and use an amplifier to sum the signals on many AT2 buses. At the IC-level, the AT2 output can be driven by a buffer amplifier with a bandwidth greater than 1 MHz, or low impedance switches can be used (e.g., 1 kΩ).

Loading Effect of AT2 Bus Capacitance on CUT   When   measuring capacitances less than 1,000 pF or resistances larger than 2 kΩ (when using an AC stimulus), the capacitive load of the AT2 bus and the voltmeter causes the impedance of the CUT to appear to be more than 1% smaller, and hence causes measurement error. The error can exceed 30% when measuring impedances comparable to the capacitance of the AT2 bus. In other words, the AT2 bus and measurement circuitry appears as an AC impedance in parallel with the CUT.

A solution to this problem is calibration (measurement) of the AT2 bus capacitance and switch impedances prior to measuring the CUT. This capacitance may then be subtracted from the measured value for the CUT to obtain the correct value. The AT2 bus capacitance is measured by connecting (using a relay) a precisely known resistance between AT1 and ground, stimulating with alternating current at the same frequency as used for the CUT, and measuring the voltage via AT2. An on-chip buffer for AT2 dramatically reduces the capacitive load connected across the CUT, but its gain must be measured.

Differences in Frequency-Attenuation for Each CUT Node   The impact of the capacitive load of the AT2 bus plus measurement circuitry also depends on the V_G switch impedance relative to the pin-AB2-AT2 switch impedance. For example, when the pin-AB2-AT2 path switch resistance is 10 kΩ, and the V_G switch impedance to V_{SS} is 1 kΩ (both values are typical), and the CUT impedance is 30 kΩ, then the apparent source impedance when measuring the higher voltage end of the CUT is 41 kΩ and the impedance when measuring the other end of the CUT (connected to V_{SS} via V_G switch) is 1 kΩ. Thus, the impact of the AT2 bus capacitance is magnified, depending on the stimulus frequency chosen.

To reduce this effect, a frequency less than 10 kHz can be used for the stimulus current. To cancel this effect, the impedances of the AB1 and V_G switches must be measured separately. The former was addressed in previous paragraphs. The latter can be measured by enabling the V_G switch at the AT1 pin (assuming it has an equal resistance), providing a stimulus current, and measuring the voltage at AT2.

Loading Effect of AT1 Bus Capacitance on Stimulus Current   The capacitive load on AT1 is just as significant as that on AT2. An off-chip bias resistance can be connected in parallel with a stimulus current (see Figure 7-5) comprising AC plus a DC offset to enable measurement of capacitors that have no parallel DC path. In this case, the stimulus is effectively a voltage source with non-zero output impedance. Any capacitance on AT1 reduces the bandwidth of the AC voltage signal reaching the AT1 pin, and appears as AC leakage current. This "leakage" current is obviously not canceled by using AC stimulus, as is DC leakage current.

Characterization of the AT1 bus path AC characteristics can address this problem. The capacitance can be measured by driving AC stimulus current into AT1, with no DC current or parallel source resistance, and measuring the resultant voltage without enabling any AT2 path.

An alternative approach is to use a frequency less than 10 kHz for the stimulus, which increases the resultant voltage and settling times.

A third approach is to use a lower value bias resistance to decrease the gain sensitivity to capacitance, however, this might require greater accuracy of the current source.

In summary, all measurements made on the 1149.4 bus must be made relative to calibration measurements. This is hardly a unique constraint - most precision meters first measure their own resistance, capacitance, inductance, or offset voltage, and many weigh scales "auto-zero" before each weighing.

Increasing the Analog Bus Bandwidth   The relatively low bandwidth of a transmission gate-based analog bus is a frequent criticism of 1149.4. Using reasonably-sized CMOS transmission gates, a bandwidth greater than 100 kHz is practical (dependent on the capacitance of the buses). While this bandwidth is practical for frequencies used by test equipment to measure passive component values, higher bandwidths are needed to accurately measure many on-chip functions. The bandwidth that is needed is clearly IC-specific, and hence is not prescribed in 1149.4. For bandwidths much higher than 100 kHz, or for accessing sensitive signals, current buffers for AT1/AB1 and voltage buffers for AB2/AT2 are essential.

When designing an IC to be 1149.4-compliant, minimizing bus wire capacitance or resistance will usually have an insignificant effect, since off-chip capacitances will be much greater – unless a voltage buffer is used in the TBIC. A more significant improvement in bandwidth is possible by reducing the switch impedance. Reducing the impedance too much, however, increases the impact on the CUT. This is especially likely when the pin being monitored is being driven by a high impedance driver, or when it is driven by an operational amplifier designed for driving only small capacitances or very high frequencies.

When designing a board for 1149.4 ICs, the AT2 bus capacitance should be minimized – it is the single factor that the board-level designer can use to improve bandwidth. The maximum on-chip total switch impedance is 10 kΩ, so to ensure a –3 dB bandwidth greater than 100 kHz, the total bus capacitance cannot exceed 66 pF. Other pins connected to AT2 will contribute to this capacitance, so reducing the number of ICs per bus helps significantly. Several board-level AT2 buses can be connected via series resistors (not necessarily all the same value) or a multiplexer to a single AT2 amplifier.

7.7.3   Noise

Aside from the systematic errors described in the preceding paragraphs, the other source of measurement error is random noise. Some noise sources are not entirely random, but appear random, such as that caused by digital transients. We shall first briefly address truly random noise. There is no specific rule in 1149.4 for permissible noise level; it is effectively addressed in the rule which requires 1% accuracy to be achieved with a 1 kHz bandwidth, while measuring a 1000 ohm resistor using a 1 kHz, 100 μA stimulus.

Random Noise   There are two types of random noise seen throughout the natural world and on ICs: thermal and 1/f noise. Other types of noise exist (e.g. shot noise), but are beyond the scope of this brief discussion – they can be addressed similarly to thermal and 1/f noise.

Thermal noise has instantaneous values which have a Gaussian distribution with a standard deviation constant across frequency. In electrical circuits, the RMS value of thermal noise increases with temperature, bandwidth, and resistance. To reduce the error caused by thermal noise, a signal can be averaged over time, at a constant temperature, in a minimum bandwidth. Noise whose amplitude is inversely proportional to frequency is called 1/f noise, and includes offset voltage. Error due to 1/f noise is reduced by measuring at higher frequencies.

Random noise can originate from the CUT if the impedances being measured are large. It can also originate from the ground reference voltage. For this reason, 1149.4 requires the reference voltage to be generated off-chip so that it is not the source of additional, unmeasurable noise. Permissible reference voltage noise is constrained by the previously-mentioned 1% accuracy rule.

Non-Random Noise   Noise sources which are not entirely random are generally caused by active digital circuitry, or by other active analog circuitry. The single most effective way to reduce the noise generated this way, is to disable any circuitry whose function is not needed to make measurements.

Active circuitry can induce noise via the substrate of an IC, via the power rails, or via capacitive coupling. Substrate coupling is minimized by careful IC design [21] and by IC wafer processing technologies such as epitaxial layers, twin wells, or insulating substrates.

Power rail noise can be caused by digital gates switching capacitive loads, and by excessive inductance or resistance in the power rail wires. Power rail noise can be transmitted to the analog bus signals via several routes. Noise on the power rail may affect the comparators or digital buffers which convert the pin voltage to a digital value in the boundary scan chain, the gates of CMOS transmission gate transistors, and the analog bus amplifiers (if present).

In extreme cases, an n-channel transistor connected to either analog bus can momentarily leak excessive current when its gate is connected to V_{SS}: If the signal connected to the transistor's source experiences a voltage glitch causing it to become sufficiently negative with respect to V_{SS}, then the transistor may turn on. This can easily occur if the source is connected to a transmission line with digital signaling present, or if an adjacent transistor switches a large current into V_{SS} and causes the local V_{SS} to temporarily increase above V_T ("ground bounce"). Similarly, p-channel transistors are vulnerable if their gates experience momentary transient voltages below V_{DD} or their sources above V_{DD}.

There are several ways to prevent these transient leakages, aside from disabling any uninvolved circuitry. Switches can be made more robust by providing better isolation between source and drain – this is done by using the "T" structure, as previously shown in Figure 7-11. The "T" structure ensures that if either of the series transistors experiences leakage current, the leakage current will go into V_{SS} (or V_{DD} as appropriate) and not into the analog bus. This technique also provides better isolation for very high frequency signals. Power rail coupling is also reduced by careful IC layout, and by off-chip power supply de-coupling capacitors.

Capacitive coupling between the buses and unrelated signals is minimized, both on-chip and off-chip, by careful layout of interconnect wires. However, some coupling is unavoidable such as when wires must cross one another. Sometimes shielding can be provided by intermediate interconnect layers. The most general way to reduce the impact of noise is to use differential signaling, as seen in Figure 7-7.

Gain It is interesting to consider the use of buffer amplifiers for AT2 which have less than unity gain. In fact, the input response range must typically include 100 mV outside the V_{SS} to V_{DD} range, and it is impossible for conventional amplifiers to drive to voltages outside the V_{SS} to V_{DD} range; therefore, the gain must be less than unity (note that CMOS transmission gates are able to convey voltages 100 mV below V_{SS} and 100 mV above V_{DD}). A lower limit to gain is not specified in the standard; it is only limited by the 1% accuracy specification. It is also worth noting that the gain may be inverting or non-inverting. In any case, the gain is usually easily addressed by the calibration procedure described previously.

7.8   Lessons from Test ICs

7.8.1   IMP (International Microelectronics Products) IC

Using a conceptual schematic designed by engineers at Hewlett Packard with suggestions from the rest of the P1149.4 Working Group, Keith Lofstrom of KLIC Ltd. designed a test chip [11] which was manufactured by IMP in 1996 and supplied free to anyone interested in evaluating P1149.4. The two most important lessons learned from that chip were:

- Current leakage can occur when the drain of a CMOS transmission gate connected to AB1 or AB2 inadvertently goes to a voltage more than 0.6 V below VSS. Such a voltage can be caused by digital signals on pins not involved in a measurement but which have an ABM. This problem and its solution are discussed in the previous section (i.e., disable digital functions and/or use T-gates).

- Transmission gates connected in series with low-impedance outputs, as a means to provide the mandatory high-impedance output state, are impractically large and affect performance dramatically. For low-impedance outputs, this function should be provided by disabling the output driver (i.e., the driver must be designed such that it can be powered down and its output not driven, a common DFT practice for digital circuits but not as common for analog).

7.8.2   Matsushita IC

With assistance from Ken Parker and John McDermid of Hewlett Packard, engineers at Matsushita Corp. designed a second test chip to evaluate P1149.4 implemented in a 0.35 μm CMOS process. They provided a selection of transistor sizes for the switches (ranging from 100 Ω to 1600 Ω), and provided 1149.1 boundary scan logic on the AT1 and AT2 pins. The important lessons learned were:

- If connections from AB1 to the function pin, and from AB2 to the function pin, are not made to the bond pad itself, then any resistance common to the two paths is added to the CUT impedance measured. In this IC the common resistance was *metal* interconnect, but it was sufficiently long and thin enough to add 43 Ω to any off-chip component measured. The correct connection method is shown in Figure 7-19C.

- Providing 1149.1 capabilities at the AT1 and AT2 pins facilitated rapid, automatic boundary scan testing of the analog bus wires.

- When V_G is equal to V_{SS}, then existing BSDL and 1149.1 test generation software can be used to generate suitable test sequences for 1149.4 ICs.

- The variation in resistance of the CMOS transmission gates used for all switches was significant, as illustrated in Figure 7-20. Lower impedance switches were provided on the IC, but the per cent variation is similar for any size.
- The transistor sub-threshold leakage was measured for 100 Ω switches (Wn/Ln=24/0.35), and when extrapolated to 1,000 transistors at maximum operating temperature with $V_T = 0.3$ V (intentionally low), was on the order of 1 mA, which is clearly unacceptable. However, large W/L ratios and low V_T are not necessary for most (if any) applications, and when they are necessary, the multiple AB1 and AB2 bus scheme described in Section 7.2 reduces leakage.
- The difference in area for a 3-bit control register and a 4-bit control register was insignificant relative to the area for a bond pad. The Working Group was very concerned about the cost of requiring several register bits at every pin, especially since 1149.1 requires only one or two. This IC greatly helped to resolve the debate about the mandatory number of ABM register bits.

7.9 Conclusions

A brief summary of 1149.4 requirements, instructions, and capabilities was presented, followed by analysis of costs and benefits, finishing with detailed discussions of many parametric issues. Addressing each of the many non-ideal characteristics of an analog bus may seem arduous, but this is the typical consideration given to most analog circuits by experienced analog designers and test engineers.

Cursory examination of the example practical schematics (Figures 7-14 through 7-16) may lead to the conclusion that the circuitry added to each analog pin is excessive, but detailed analysis of the requirements reveals that the circuitry is minimal.

For fast, simple interconnect testing using 1149.1, two register bits are needed to control the 3 digital states (0, 1, Z); and one additional state (a reference voltage) is required by 1149.4. A third bit controls access to an analog stimulus (AT1), and the fourth bit controls access to the monitoring bus (AT2). All combinations of digital states, analog stimulus, and analog observation have applications which have been identified. It is difficult to imagine using less circuitry without significantly compromising test capability.

ShiftDR and ClockDR are essential for sampling in parallel and shifting serially, and UpdateDR is essential for applying stimulus without glitches. Mode1 and Mode2 enable quick, reliable transition between four essential modes: mission (*BYPASS*, etc.); mission plus limited test (*PROBE*, *INTEST*); 1149.x device test mode (*EXTEST*, etc.); and non-1149.x test mode (*HIGHZ*).

The unique TBIC (with optional, differential ATAP) allows hierarchical, multi-bus architectures which can be calibrated for continuously improving accuracy, and which should accommodate future growth in IC complexity along with decreasing voltage supplies.

The facilities listed above comprise a minimal set, yet this minimal set allows a mixed-signal tester per pin. Its only limits are the creativity of the world's designers and test engineers.

References

1. P. Fasang, D. Mullins, and T. Wong, "Design for Testability for Mixed Analog/Digital ASICs," Proceedings of the *IEEE Custom Integrated Circuits Conference*, pp. 16.5.1-16.5.4, May 1988.
2. K. Wagner and T. Williams, "Design for Testability of Mixed Signal Integrated Circuits," Proc. of *IEEE International Test Conference*, pp. 823-828, October 1988.
3. Minutes of P1149.4 Working Group Meeting, Seattle, July 16-17, 1992.
4. "Panel: P1149.4 Mixed-Signal Test Bus Framework Proposals," Proc. of *IEEE International Test Conference*, pp. 554-557, September 1992.
5. M. Jarwala, "Design for Test Approaches to Mixed-Signal Testing," Proc. of *IEEE International Test Conference*, pp. 555, September 1992.
6. "Panel: Mixed-Signal Test Bus: Has It Arrived?" Proc. of *IEEE International Test Conference*, pp. 590-592, October 1993.
7. K. Parker, J. McDermid, S. Oresjo, "Structure and Metrology for an Analog Testability Bus," Proc. of *IEEE International Test Conference*, pp. 309-322, October 1993.
8. C. Thatcher, R. Tulloss, "Towards a Test Standard for Board and System Level Mixed-Signal Interconnects," Proc. of *IEEE International Test Conference*, pp. 300-308, October 1993.
9. J. Matos, A. Leão, J. Ferreira, "Control and Observation of Analog Nodes in Mixed-Signal Boards," Proc. of *IEEE International Test Conference*, pp. 323-331, October 1993.
10. S. Sunter, "The P1149.4 Mixed-Signal Test Bus: Costs and Benefits," Proc. of *IEEE International Test Conference*, pp. 444-450, October 1995.
11. K. Lofstrom, "A Demonstration IC for the P1149.4 Mixed-Signal Test Standard," Proc. of *IEEE International Test Conference*, pp. 92-98, October 1996.
12. S. Sunter, "Cost/Benefit Analysis of the P1149.4 Mixed Signal Test Bus," IEE Proc. – Circuits, Devices and Systems, Vol. 143, No. 6, pp. 393-398, December 1996.
13. "IEEE Standard for a Mixed-Signal Test Bus," (Draft 18), The IEEE, Inc, 345 East 47th St., New York, NY, June 1997.

High Frequency Analog Bus Methods

14. P.M. Van PeteGhem and J.F. Duque-Carrillo, "Compact High-Frequency Output Buffer for Testing of Analog CMOS VLSI Circuits," *IEEE Journal of Solid-State Circuits*, vol. 24, pp. 540-542, April 1989.
15. S. Sunter, "A Low Cost 100 MHz Analog Test Bus," Proceedings of *IEEE VLSI Test Symposium*, pp. 60-65, May 1995.
16. R. Mason, B. Simon, K. Runtz, "On-Chip Analog Signal Testing Using an Undersampling Approach," Informal Proc. of *International Mixed-Signal Testing Workshop*, pp. 165-172, May 1996.
17. Minutes of P1149.4 Working Group Meeting, Washington DC, October 1996.
18. K. Lofstrom, "Early Capture for Boundary Scan Timing Measurements," Proc. of *IEEE International Test Conference*, pp. 417-422, October 1996.

1149.1

19. A. Cron, "IEEE-1149.1 Use in Design For Verification and Testability at Texas Instruments," Proc. of *IEEE ASIC Seminar and Exhibit*, pp. P4_1.1-P4_1.5, 1989.

20. "IEEE Standard Test Access Port and Boundary-Scan Architecture," IEEE Standard 1149.1-1990 (Includes IEEE Std 1149.1a-1993), The IEEE, Inc, 345 East 47th St., New York, NY, October 1993.

Addressing Noise in Mixed-Signal ICs

21. B. Stanisic, N. Verghese, R. Rutenbar, R. Carley, D. Allstot, "Addressing Substrate Coupling in Mixed-Mode ICs: Simulation and Power Distribution Synthesis," *IEEE Journal of Solid-State Circuits*, vol. 29, pp. 226-238, March 1994.

CHAPTER 8

Test Techniques for CMOS Switched-Current Circuits

Chin-Long Wey

$\mathbf{A}$ class of analog circuits wherein current rather than voltage is the primary signal medium has recently received considerable attention. Current-mode circuits offer a potential for significant speed improvement. Further, low-cost CMOS processes used for digital circuits can be used to fabricate analog circuits using the switched-current (SI) technique. Thus, they are ideal for mixed-signal applications. This has led to the development and implementation of high-performance CMOS switched-current circuits for low-power/low-voltage signal processing applications. Testing such high-performance circuits in low-power/low-voltage environments is a challenging task. This article reviews existing test methods developed for switched-current circuits and addresses some important issues for future research.

8.1 Introduction

In mixed-signal circuits with a small analog interface component, using the same supply-voltage for both the analog and digital components reduces overall system cost by eliminating the need to generate multiple supply voltages with dc-dc converters. Therefore, to be compatible with low-voltage systems, analog signal processing components must be able to operate at supply voltages in the 2-3 V range. Reducing the power dissipation associated with high-speed sampling and quantization is another important factor.

Traditionally, the switched-capacitor (SC) technique [4] has been employed extensively in the analog interface portion of mixed-signal designs. However, SC circuits are not fully compatible with digital CMOS processing technology and as the technology advances further the drawbacks of SC technique are becoming more significant [5]. Switched-capacitor circuits traditionally require high quality linear capacitors, which are usually implemented using two layers of polysilicon. The second layer of polysilicon used by switched capacitors is not needed by purely digital circuits and may become unavailable as process dimensions shrink to the deep submicron range. The trend towards submicron processes is also leading to a reduction in supply voltages, directly reducing the maximum voltage swing available to SCs and consequently reducing their maximum achievable dynamic range. With lower supply voltages the realization of high-speed high-gain operation amplifiers become more difficult [5].

Recently, a class of analog circuits wherein current rather than voltage is the primary signal medium has received considerable attention [6-21]. The use of the current-mode as the signal medium creates a potential for speed improvement because stray-inductance effects in such low-impedance switched-current (SI) circuits are much less severe than in high-impedance SC circuits. The SI technique couples itself well with the down-scaled CMOS technology, where transistors with a high cut-off frequency are available, meaning a high calibration speed. In addition, with the SI technique, highly-linear capacitances are not needed to achieve high-accuracy analog signal processing. Thus, with the SI technique, the same low-cost digital CMOS process used for the digital component of mixed-signal circuits can also be used for the analog component. This has led to the development and implementation of high-performance CMOS switched-current circuits for low-power/low-voltage signal processing applications.

Analog MOS circuits are becoming increasingly sophisticated in terms of checking and correcting themselves [22]. Self-correcting, self-compensating, or self-calibrating techniques eliminate errors traditionally associated with analog circuits. They have been adopted to eliminate offset and nonlinearities, and cancel the errors produced by such effects [23]. Self-compensating/calibration techniques may work properly when some component values deviate from their nominal values within a certain tolerable percentage. However, the self-compensating/calibration techniques may no longer work properly in the presence of faulty switching elements. Testing high-performance SI circuits in a low-power/low-voltage environment is a challenging task [25-43]. This chapter reviews current work in test techniques for SI circuits.

Since the current copier is the major building block of SI circuits, an SI circuit can be easily tested if the copiers it employs are fully testable. The questions are: *how can one design testable current copiers?* and *how can one efficiently generate test sets for them*? The function of a current copier can be evaluated by comparing the stored current with the (original) input current. Since the output current from the copier has a polarity opposite that of the input current, simply adding the output current in the following clock cycle to the original input current results in a current difference that directly indicates the presence of current-copying errors and faults in the copier. This has motivated the development of self-testing designs for current copiers [36-39], a divide-by-two circuit [36], and integrator based biquadratic filters [37,38]. In order to enhance testability and diagnosability of analog circuits, a scan structure using an analog sample-and-hold (S/H) circuit as an analog shift register was developed [44] for voltage measurement data. Scan structures using current copiers as registers have also been implemented [41-43].

To enhance the reliability of analog-to-digital (A/D) converters for real-time applications, a CMOS current-mode algorithmic A/D converter which possesses the concurrent error detection (CED) capability was developed in [25-30]. The CED scheme detects both transient and permanent faults. The potential faulty behaviors of the converter circuit for each possible single switch fault has been studied. The fault model assumes a single faulty switch is either permanently in the stuck-at-ON state (*S/ON*) or the stuck-at-OFF state (*S/OFF*). The failure of components other than switches can be modeled as faults in the switches associated with the faulty component. The converter is fully testable for the single stuck-at fault model. Due to the simplicity of the test patterns generated for the A/D converter [25-30], a built-in self-test (BIST) design, with a simple structure, has been developed and implemented in [30]. The test generation process and BIST design concept have been extended to both catastrophic and parametric fault models [32-35]. In this effort macromodels of faulty switches were developed for test generation and fault simulation of SI circuits.

This chapter is structured as follows. In Section 8.2, the structures of current copiers and their operation are reviewed. The review is followed by a presentation of test generation techniques for current copiers [37,38], a self-testing current copier design [36], and a built-in current tester design [33,39]. Section 8.3 discusses the design and operation of a current-mode CMOS algorithmic A/D converter [7] and presents test generation techniques, a concurrent error detection scheme, and a BIST design [25-34] for that A/D converter. Section 8.4 discusses SI scan structures and a design for test technique comparable to digital scan. The chapter ends with concluding remarks and a discussion of future research directions in Section 8.5.

8.2 Current Copiers: Basic Building Blocks of SI Circuits

The switched-current (SI) technique is a current-mode signal processing technique, where current-mode signal processing is defined as analog signal processing in which current rather than voltage is the primary information-carrying medium. The current copier is the basic building block of SI circuits, and the performance of an SI circuit is generally determined by that of the current copiers it employs. This section reviews the structures of current copiers, their operation, and the existing methods for testing current copiers and SI circuits.

8.2.1   Structure and Operation

A current copier, shown in Figure 8-1(a), is comprised of switches S_1 and S_2, the current-storage transistor M_1, and the holding capacitor C. To copy the current I_{in}, switches S_1 and S_2 are turned on feeding the current I_{in} to M_1 and C. The capacitor is charged to the gate voltage needed by M_1 to support a current equal to I_{in}. When both S_1 and S_2 are off, the copier cell is disconnected from the current source; thereafter the copier cell is capable of sinking a current I_{in} when connected to a load. Thus, the current copier can reproduce the input current I_{in} without requiring well-matched elements. However, the copier suffers from two major error effects due to (1) nonzero conductance of M_1 and (2) charge-feedthrough of S_2 [5-7]. The former results from the channel length-modulation effect and the drain-gate capacitive coupling of M_1. The latter error effect is caused by the charge stored in the conducting channel of a MOS transistor.

The error effect due to nonzero conductance of M_1 can be alleviated using a negative feedback structure [5], as shown in Figure 8-1(b), where an amplifier is inserted between the drain and the gate of M_1. Since the transistor M_1 and the resistance of the input current source together constitute a voltage inverter, referred to as the *inverter-negative-feedback copier*, a positive-gain amplifier is needed to keep the feedback negative. To resolve the instability problem of the inverter-negative-feedback copier, a follower-negative-feedback copier, as illustrated in Figure 8-1(c), was developed [17-19]. The copier achieves a dynamic range from 300 µA to 550 µA with a settling time of 3 ns, for a 3.3 V supply voltage and a 2 µm CMOS process[19]. The charge-feedthrough error effect can be reduced by either increasing the capacitance C, or using appropriate switches. Many other current copiers, such as cascode, regulated cascode, and S^2I copiers have also been developed. The detailed information about these copiers can be found in [5].

Switches can be implemented by NMOS, PMOS, or CMOS transistors. The use of the NMOS transistor as the switch may suffer from the following two error effects: (1) the threshold voltage of the transistor may clamp the input voltage signal and cause signal distortion; and (2) when the switch is connected to a capacitor C, turning off the switch will force the charge stored in the switch to dump to C and cause an error, referred to as *charge-feedthrough error*. The use of a CMOS switch with a dummy switch, as shown in Figure 8-1(d), can alleviate the charge-feedthrough error. More specifically, the CMOS switch, formed by transistors M_1 and M_2, can eliminate the voltage swing limit clamped by the threshold voltage. The dummy switch, constructed by transistors m_3 and m_4, is half the size of the CMOS switch and produces a gate capacitance which is half that of the CMOS switch.

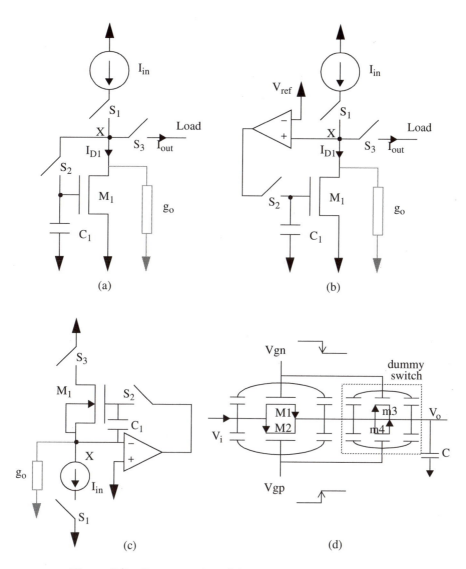

Figure 8-1 *Current copiers: (a) simple version; (b) negative feedback approach; (c) alternative negative feedback approach; (d) switch. (Figure courtesy of the IEEE, © 1996 [19].)*

Therefore, $C_{gs1} = C_{gs4} + C_{gd4}$ and $C_{gd2} = C_{gs3} + C_{gd3}$. Turning off the switch will force V_{gn} to low and V_{gp} to go high. Thus, the charge absorbed by C_{gs1} is offered by the gate capacitance of m_4, while the charge emitted from M_2 will be absorbed by the gate capacitance of m_3. The charge stored in C can remain the same when the switch is turned off.

The current copiers in Figure 8-1 are implemented with two types of switches: *current switches* and *voltage switches*. The former have been commonly used in switched-current circuits, while the latter have been used in both switched-capacitor and switched-current circuits. As their names imply, the current switch passes current as a signal, while the voltage switch passes voltage as a signal. For example, in Figure 8-1(a), S_3 is used to pass the currents held in the copier and S_1 passes the input current I_{IN}. Thus, they are implemented with current switches. On the other hand, S_2, used in the copier for calibration, is implemented with voltage switches. It has been shown that the effect due to charge-feedthrough error is not significant in current switches, but it causes a significant fault in voltage switches [32-35].

8.2.2   Testing Current Copiers

The basis of any switched current-mode environment is essentially the current copiers and little additional test circuitry is required [37,38]. Figure 8-2(a) shows a typical second generation cascode current copier. The principle of the test technique relies essentially on the principle of its operation: that of charge storage and transportation from the gate source capacitance of the first memory transistor M_1 to the next in the circuit. Each memory transistor is biased with M_2 and M_3, and switched to subsequent cells with M_4 and M_5. Since SI circuits are essentially sequential, a fault exhibited anywhere in the circuit should be carried sequentially through the circuit and hence should be observable at the output [38].

Based on the transistor level fault models, such as the gate-source short (GSS), gate-drain short (GDS), drain open (DOP), and source open (SOP), several test methods have been investigated [37,38], including (1) pre/post-subtraction digitization; (2) linear tolerance band definition; (3) frequency domain variance; and (4) histographic analysis. Linear tolerance provides a readily practicable implementation of fault detection, while histographic analysis is a simple efficient method for the comparison of test vectors. The study discussed the effect of test stimuli for various catastrophic faults and demonstrates the fault detectability. Results show that the cascades of current copiers are highly testable with standard test patterns [38].

Switched-current circuits consist entirely of current copiers with feedback loops and some additional (biasing) transistors providing specific functionality. Since the current copiers are highly testable, there is a clear immediate gain in testability if global feedback loops can be disconnected for test. Rather than comparing the signatures, the outputs of two equal length cascades of copiers on the same circuit can be compared with one another as shown schematically in Figure 8-2(b) [38]. In other words, the functional circuit is reconfigured to provide two subcircuits. Consider an integrator based biquadratic filter. Modifying the circuit by add-

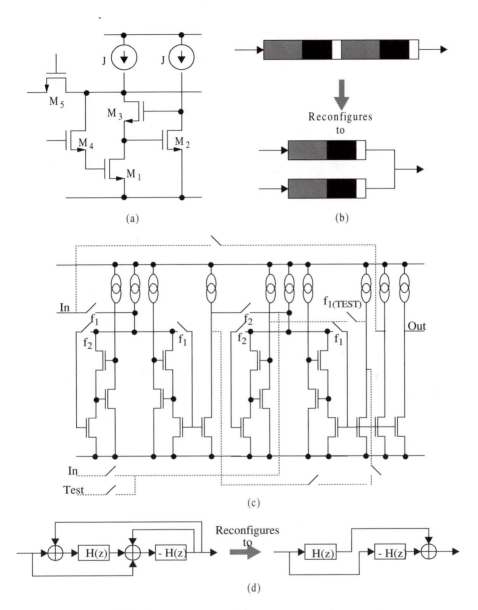

(a)

(b)

(c)

(d)

Figure 8-2   *Test methods: (a) cascode current copier; (b) circuit reconfiguration; (c) reconfigurable integrator-based biquad; (d) configuration. (Figure courtesy of the IEE, © 1993 [38].)*

ing five extra in-circuit test switches, as shown in Figure 8-2(c), gives self-testing capability. Block diagrams of the filter in normal and test modes are illustrated in Figure 8-2(d). The circuit functions as either a biquadratic filter section or a self-testing integrator section, depending upon the settling of the test switches. The two integrators are reconfigured to provide equal and opposite current levels at their outputs. In the test mode, the sum of their outputs should be zero.

Based on similar fault models, self-testing designs for current copiers and switched-current building blocks were developed in [36]. Basically, a current copier can evaluate its function by comparing its stored current with the (original) input current. The output current has a polarity opposite that of the input current. Thus, simply adding the output current to the original input current (in the subsequent clock cycle), as shown in Figure 8-3, results in a current difference that directly indicates the presence of current-copying errors and faults in the copier. The test cir-

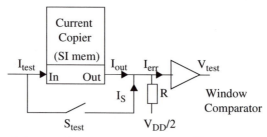

Figure 8-3 *A self-testing current copier. (Figure courtesy of the IEEE, © 1996 [36].)*

cuitry for the self-test design includes a switch S_{test} to connect the input to the output node, a test-current generator, and a window comparator to detect the error current thresholds. The input current I_{test} is near full-scale but does not require a high absolute accuracy because only current differences are processed. The only requirement is that the input current must remain constant for two clock cycles. For catastrophic faults, the current comparator does not need to be very accurate to achieve a good fault coverage. The test circuitry produces a go/no-go signal easily accessed via a standardized scan path (IEEE std 1149.1). Several undetectable faults have been analyzed with a regulated cascode copier, and alternative test methods have been evaluated for improving testability.

To simplify the test generation process for current copiers and SI circuits for both catastrophic and parametric faults, a high-accuracy current comparator, shown in Figure 8-4(a) [33,39], was developed as a built-in tester that checks if the difference between two input currents I_{i1} and I_{i2} is sufficiently small. With high accuracy but moderate speed, the tester can be operated in a low-voltage/low-power environment. The tester is comprised of a current comparator, a voltage window comparator, and a digital latch. Assume the comparator is designed such that for any input

$I_x = I_{i1} - I_{i2}, -I_{to1} \le I_x \le I_{t01}$, a proportional voltage level $V_y, -V_w \le V_y \le V_w$, is generated. Then, a simple voltage window comparator with a pair of symmetric threshold voltages, V_w and $-V_w$, can be used for current comparison. An ideal I-V current comparator, has a linear relationship $V_y = I_x \times r_k$, where r_k is a constant transresistance.

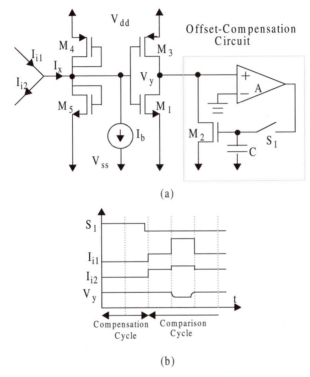

(a)

(b)

Figure 8-4 *Built-in tester: (a) schematic; (b) switching sequence.*

In practice however, the accuracy may be affected by: (1) the offset current due to mismatched components in the current comparator; and (2) the nonlinearity of r_k. The offset current may cause an offset voltage, V_{ofs}, in the output of the current comparator. The nonlinearity of r_k leads to a nonlinearity component $Vnl(I_x)$, which is a function of I_x, in the output voltage. Thus, the output voltage $V_y = I_x \times r_k + V_{ofs} + V_{n1}(I_x)$. The accuracy of the current comparator can be improved by reducing the terms V_{ofs} and $V_{nl}(I_x)$. The circuit has been designed and simulated with the *MOSIS SCN* 2 μm CMOS process parameters and 2 V supply voltage. The current copier has a simple structure and its layout area is 0.01 mm². Simulation results show that the proposed circuit achieves a small offset current, 0.1

nA, and low power dissipation, 20 μW. For design simplicity, a voltage window comparator with symmetric threshold voltages is used, where the threshold voltage is determined by the testing resolution and can be used to compute the testing confidence.

SI circuits consist entirely of current copiers with feedback loops and some additional (biasing) transistors providing specific functionality. If the current copiers can be fully and individually tested by the built-in tester, then the testability of SI circuits under test will be determined by that of switches on the global feedback loops.

8.3 Testing of Switched-Current Algorithmic A/D Converters

Although mismatched components are allowed in current copiers, Copiers are still susceptible to faulty switching elements. A faulty switching element may cause the copier to hold an incorrect current. This section reviews the design and operation of an algorithmic CMOS current-mode A/D converter [7], and presents a concurrent error detection scheme, test generation techniques, and a built-in self-test (BIST) design for the converter.

8.3.1   Structure and Operation

Figure 8-5(a) illustrates an algorithmic A/D converter that combines current mode and dynamic techniques [7]. The converter does not rely on high gain amplifiers or well-matched components to achieve high resolution and is inherently insensitive to the amplifier's offset voltage. The converter is comprised of two NMOS current copiers, one PMOS copier, an op-amp, and a current comparator. For an N-bit digital output, the operation is as follows:

1. The converter starts converting for the *most significant bit* (MSB) of an input current I_{IN} by first switching on S_1, S_2, and S_3. This causes the current in N_1 to be set to I_{IN}.
2. Once the current in N_1 is held, S_2 and S_3 are switched off while S_4 and S_5 are on to copy I_{IN} to N_2.
3. Once the input signal has been stored in N_1 and N_2, twice the input signal is loaded into P_1 by turning off S_1 and S_5 while switching on S_2, S_6, and S_7.
4. After P_1 is set, S_2, S_4, and S_7 are turned off while S_{10} is turned on, thus allowing the comparator to sense the current imbalance, and hence, determine if the signal, $2I_{IN}$, is greater than I_{ref}.
5. If the signal exceeds the reference, the MSB will be a "1," otherwise it will be a "0." This completes the conversion for the MSB.
6. The remaining (N-1) bits are then converted in the same manner. The signal held in P_1 is loaded to N_1 by turning on S_6, S_2, and S_3. If the preceding bit was

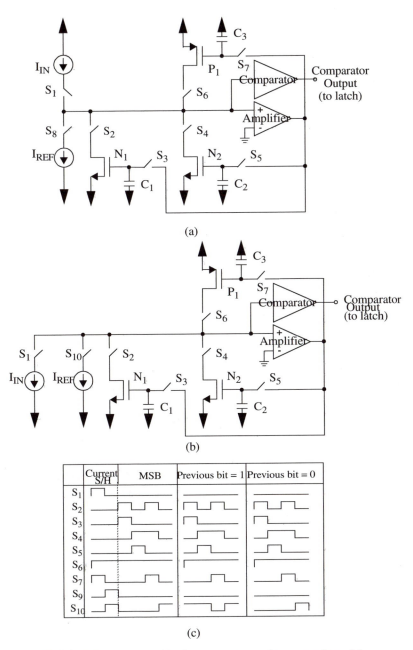

Figure 8-5 *An algorithmic current mode converter: (a) schematic [7]; (b) schematic [30]; (c) switching sequence. (Figure courtesy of the IEEE, © 1995.)*

a "1," S_{10} is also turned on to subtract the reference from the signal in P_1. On the other hand, if it was a "0," S_{10} is off so that the signal remains unchanged.

7. Once N_1 is set, N_2 is set by the same procedure. The signal is then doubled and stored on the gate of P_1. Finally, it is compared with the reference.

8. This sequence is repeated until the desired resolution has been achieved; an end of conversion pulse is then generated to signal the end of conversion. (The detailed switching operation can be found in [7].

The converter needs 4N clock cycles for an N-bit data conversion. The proto-type circuit has been fabricated using a 3-μm CMOS technology and achieved a res-olution of 10 bits with a maximal sampling rate of 25 KHz, or a 40 μsec conversion time. The power supply was +5 V and the reference current was 100 μA. It should be mentioned that, since the input current is needed during the first two clock cycles for converting the MSB, a sample-and-hold (S/H) circuit is required for the input current. However, the S/H circuit can be omitted by holding the input I_{IN} in P_1, where the polarity of the input current is changed as shown in Figure 8-5(b) [28-30]. Figure 8-5(c) illustrates the switching sequence for the modified converter.

8.3.2 Concurrent Error Detection (CED)

Although reliability may be improved by using sophisticated testing schemes to weed out faulty components, such off-line or static tests cannot identify transient faults that occur during real-time operation. The term transient fault, or temporary fault, refers to faults for which the duration of fault behavior is relatively short. The term permanent fault refers to faults for which the duration of the fault behavior is sufficiently long. Transient faults are very common in today's digital VLSI design. Since all switches in the A/D converter are controlled by the digital clock signals, a signal may temporarily change its value from 0 to 1 or from 1 to 0 and cause the switching elements to temporarily malfunction. It would be preferable for the cir-cuits to be designed such that they indicate any malfunction during normal opera-tion. That is, a circuit should not produce erroneous results without an indication to the outside world. A mechanism for *concurrent error detection* (CED) must be installed to detect transient faults before they produce undesirable results. In order to enhance the reliability of A/D converters for real-time applications, an alternative current-mode A/D converter which possesses the ability to concurrently detect tran-sient faults and permanent faults during normal operation, has been developed [25-30].

Figure 8-6(a) illustrates a CED scheme with an alternating logic (AL) imple-mentation. First, the input current $I_{t1} = I_{IN}$ is converted during the first time step (or, normal operation phase that consists of several clock cycles) and the resulting digital data is stored in a digital shift register. Then, the complemented current

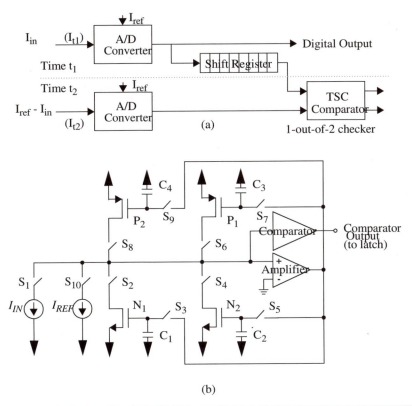

(a)

(b)

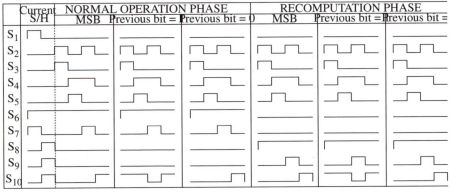

(c)

Figure 8-6 *Concurrent error detection: (a) with AL implementation; (b) schematic; (c) switching sequence. (Figure courtesy of the IEEE, © 1995 [30].)*

$I_{t2} = I_{ref} - I_{IN}$ is converted during the second time step (or recomputing phase). The digital data resulting from both phases are compared to detect any errors that may exist. If the converter is fault-free, the converted data resulting from both phases must be bitwise (one's) complements of each other. For example, with the reference current suggested in [8], i.e., $I_{ref} = 100$ μA, the input current 27 μA and its complement 73 μA are converted to the 10-bit data $D_1 = (0100010100)$ and $D_2 = (1011101011)$, respectively. As expected, D_1 and D_2 are bitwise complements. Since the comparison is in a digital manner, a *totally self-checking (TSC) checker* can be used to identify the error and also to ensure the correctness of the checker circuit. This ensures that data has been reliably converted and is error free.

Figure 8-6(b) shows the current-mode A/D converter with the CED capability [25-30]. The input current I_{IN} is sampled only once and the current is stored in P_1. In order to store the current $(I_{ref} - I_{IN})$, an additional PMOS current copier is needed to hold the current at the beginning of the data conversion. The input current I_{IN} is copied and stored in P_1 by turning on switches S_1, S_6, and S_7, while the current difference $(I_{ref} - I_{IN})$ is loaded to P_2 by turning off S_1 and S_7 and turning on S_8, S_9, and S_{10}. Once both currents are stored, the currents held in P_1 and P_2 are converted. The results from both conversions are compared to identify an error, if it exists. The data conversion process is exactly the same as presented previously. Figure 8-6(c) illustrates the switching sequence.

8.3.2.1   Fault Model and Faulty Behavior

Although mismatched components are allowed in the converter of Figure 8-5(a), the converter is still susceptible to faulty switching elements. Any faulty switching element may result in errors in the converted data. The single stuck-at fault model has been commonly employed for digital test generation. We extend it to the algorithmic converter. In this implementation, it is assumed that only one faulty switch occurs at a time and the faulty switch is permanently stuck-at either the ON state (*S/ON*) or the OFF state (*S/OFF*). In general, the faults may be caused by either a malfunctioning clock generator, i.e., one bit may be permanently stuck-at-1 (or 0) causing the controlled switch to be *S/ON* (or *S/OFF*), or by malfunctioning transistor switches. Although the fault model considered here focuses on the faulty switching elements, faults that occur in other elements can be modeled by equivalent faults in the switching elements. For example, consider the current copier for the transistor N_2 and the capacitor C_2. A short between the drain and the source, or the gate and the source, of N_2, is equivalent to an *S/OFF* at S_4 or S_5. (The short may be caused by metal lines across the terminals); a short between the drain and the gate is equivalent to an *S/ON* at S_5; an open at the drain or source is equivalent to an *S/OFF* at S_4.

In general, the duration of a transient fault is likely to be shorter than the conversion time for the converter, i.e., 4N clock cycles. Here, a fault may occur during the first time step, or during the second time step, or overlap in both time steps (but with a duration shorter than 4N). If the fault occurs only during the first time step, i.e., the fault disappears during the second time step, then the converted data D_2 is reliable and guaranteed to be error-free. D_2 can be used to check D_1 for potential errors. Errors in D_1 will be detected by the comparison process. Thus, the fault is detectable. Similarly, a fault that occurs only during the second time step is also detectable. Now, if the fault occurs after the r-th bit of D_1 is converted and disappears after the (r-1)-th of the D_2, for any integer r, then at least the first (r-1) bits of D_1 are error-free and can be used to detect the fault. Thus, the fault is detectable and the proposed design can detect all transient faults.

If the duration of transient fault is longer than the time required to complete the first time step and the second time step, all time redundancy CED schemes will no longer possess the property of disjoint error sets, and the errors cannot be detected. This implies that not all permanent faults can be detected by this CED scheme. (Note that permanent faults here mean that the switching element is permanently stuck-at *ON/OFF* during the normal operation.) It would be preferable for the circuits to be designed such that they will indicate any malfunction during normal operation and will not produce an erroneous result without an error indication. That is, it is preferable that they be *fault secure*. Informally, in such circuits, any errors produced by a fault should be detected by an error detection mechanism and their presence should be indicated by an error signal. No fault should be allowed to produce errors that are not flagged by the error signal.

Table 8-1 lists the detection status for all possible stuck-at faults that may occur at the switching elements in the converter. The errors that can be definitely detected by the CED scheme are referred to as *Type 1 errors*, i.e., a Type 1 error causes $D_1 \neq \overline{D}_2$ for all possible input currents. The errors that cannot be detected are referred to as *Type 2 errors*, i.e., a Type 2 error causes $D_1 = \overline{D}_2$ for all possible input currents. In some cases $D_1 \neq \overline{D}_2$ for all possible input currents except a few. For these few where $D_1 = \overline{D}_2$, if the resulting data D_1 is reliable even in the presence of fault(s), then the circuit is fault-secure and such an error is referred to as a *Type 3 error*. On the other hand, if the resulting data D_1 is not reliable in the presence of fault(s), then the fault cannot be detected for the application of such input currents and this ia a *Type 4 error*.

8.3.2.2   Fault Coverage

Table 8-2 summarizes the effects of the above mentioned errors. According to Table 8-1, there exist eight Type 1 errors, four Type 2 errors, five Type 3 errors, and three Type 4 errors. If the fault coverage is defined as the total number of Types 1, 3, and 4 errors over all possible errors, the fault coverage of permanent faults is 80%

Table 8-1   *Error Effects. (Table courtesy of the IEEE, © 1995.)*

Type 1 Errors: $D_1 \neq \overline{D}_2 \; \forall \; I_{IN}$: Definitely detectable.

Fault	Conditions	Fault Effects
$S_2(S_4)$ S/OFF	Zero current held in N_1 (N_2). Current not doubled.	$D_1=D_2=(00...00)$
$S_2(S_4)$ S/ON	N_1 (N_2) cancels current from P_1 or P_2. N_2 (N_1) holds zero current. Current not doubled.	$D_1=D_2=(00...00)$
$S_3(S_5)$ S/ON	Output depends on the CMOS structure P_1/N_1 (P_1/N_2)	A bit in D_1 equals the corresponding in D_2.

Type 2 Errors: $D_1 = \overline{D}_2 \; \forall \; I_{IN}$: Definitely undetectable.

Fault	Conditions	Fault Effects
S_1 S/OFF	Equivalent to $I_{IN} = 0$. P_1 holds zero and P_2 holds I_{ref}.	$D_1=\overline{D}_2=(00...00)$
S_6 S/OFF	P_1 never copies current. P_2 holds I_{ref}.	$D_1=\overline{D}_2=(00...00)$
S_{10} S/OFF	Equivalent to $I_{ref} = 0$ - I_{IN} forced into P_2.	$D_1=\overline{D}_2=(11...11)$
S_{10} S/ON	I_{ref} always added to the current copied into P_1. $-I_{IN}$ forced into P_2.	$D_1=\overline{D}_2=(11...11)$

Type 3 Errors: D_1 correct or $D_1 \neq \overline{D}_2 \Rightarrow$ Fault secure.

Fault	Conditions	Fault Effects
S_6 S/ON	Normal operation phase not altered. Residual current in P_1 always sourced during recomputation phase.	D_1 correct D_2 random
S_7 S/ON	During comparison P_1 initially copies I_{ref} and then gets compared. Recomputation phase not altered.	$D_1=(11...11)$ D_2 correct
S_8 S/ON	Residual current in P_2 always sourced during recomputation phase. Recomputation phase not altered.	D_1 random D_2 correct
S_9 S/ON	Normal operation phase not altered. During comparison P_2 copies I_{ref} and then gets compared.	D_1 correct $D_2=(11...11)$

Type 3 Errors: D_1 correct or $D_1 \neq \overline{D}_2 \Rightarrow$ Fault secure.

Fault	Conditions	Fault Effects
S_8 S/OFF	Normal operation phase not altered. P_2 never copies current.	D_1 correct $D_2=(00...00)$
S_9 S/ON	Normal operation phase not altered. Residual current in P_2 always compared with I_{ref}.	D_1 correct D_2 random

Type 4 Errors: Random data $\Rightarrow$ Most likely detectable.

Fault	Conditions	Fault Effects
S_1 S/ON	Varying I_{IN} always sourced to the circuit.	D_1 and D_2 random
$S_3(S_5)$ S/OFF	N_1 (N_2) do not copy any current but its residual current is sourced.	D_1 and D_2 random
S_7 S/OFF	Residual current in P_1 always compared with I_{ref}. Recomputation phase converts the complement of this residual current.	D_1 and D_2 random

Table 8-2   *Error Detection. (Table courtesy of the IEEE, © 1995.)*

Switches	S/ON	S/OFF	Switches	S/ON	S/OFF
S_1	4	2	S_6	3	2
S_2	1	1	S_7	3	4
S_3	1	4	S_8	3	3
S_4	1	1	S_9	3	3
S_5	1	4	S_{10}	2	2

8.3.3   Test generation

Similar to the fault model and fault behaviors discussed in Section 8.3.2.1, a single stuck-at fault model is also considered in the test generation process. At most one faulty switch is assumed to be present in the circuit at any instant and a faulty switch may be permanently or temporarily *S/ON* or *S/OFF*.

8.3.3.1   Fault Behaviors

The analysis of the faulty switches in the converter of Figure 8-5(b) revealed that, due to the fault equivalence, potential faulty behaviors can be classified into three types: *Type 1 fault* behavior occurs when the faulty switch results in the same conversion output regardless of the values of the input current. Switches S_1, S_2, S_4, S_7, and S_{10} being *S/ON* and S_1, S_2, S_4, S_6, and S_{10} being *S/OFF* are representative of this fault behavior; *Type 2 fault* behavior occurs when the faulty switch renders the conversion output dependent on the initial condition of the active capacitors. Switches S_3, S_5, and S_7 lead to this condition when *S/OFF*; and *Type 3 fault* behavior makes the result of the conversion process dependent on the CMOS structure P_1/N_1 (or P_1/N_2) when S_3 (or S_5) is being *S/ON*. Throughout the next analysis, I_{P1} (I_{N1} or I_{N2}) will denote the current held in P_1 (N_1 or N_2).

More specifically, consider the case when S_1 or S_6 is *S/OFF*, the input current will not be copied into P_1; this is effectively equivalent to an input current of zero. S_{10} being *S/OFF* leads to the current in P_1 being compared to zero instead of I_{ref}. Hence conversion results in a string of ones. For simplicity, Type 1 faulty elements are distinguished as *Type 1A* if they result in a string of zeros, and as *Type 1B* if they produce a string of ones. The detailed fault behaviors for various fault types can be found in [30].

8.3.3.2   Test Generation and Fault Coverage

Figure 8-7(a) summarizes the expected output bit string for each type of faults. A string of zeros is expected in the presence of a Type 1A fault for any input current applied to the converter. Similarly, a string of ones is expected for Type 1B fault. Thus, two test currents, $I_{T1} = 0$ and $I_{T2} = I_{ref}$, can detect both types of faults. More specifically, a bit string of zeros is expected when the test current $I_{T1} = 0$ is applied to a fault-free converter. Thus, the test current $I_{T1} = 0$ detects Type 1B faults. Similarly, the expected bit string of ones for $I_{T2} = I_{ref}$ detects Type 1A faults. For Type 2 faults, the application of $I_{IN} = 0$ results in a bit-string pattern of 0.01x..x. This implies that the expected bit string of zeros for the test pattern $I_{T1} = 0$ detects the faults. On the other hand, a bit pattern of 1..10x..x is produced when $I_{IN} = I_{ref}$ is applied. Thus, the expected bit string of ones for the test current $I_{T2} = I_{ref}$ detects such faults. Finally, in the presence of a Type 3 fault, a bit string containing at least a single "1" is generated and the fault can be detected by the test current $I_{T1} = 0$. This implies that both test currents, $I_{T1} = 0$ and $I_{T2} = I_{ref}$, detect all single stuck-at faults at the switching elements of the converter in Figure 8-5(b). If the *fault coverage* is defined as the ratio of the number of faults can be detected over the total number of faults in that circuit, then the converter is fully testable with respect to single switching element faults using these two test currents.

8.3.4   BIST Design

Since most switched-current circuits in mixed-signal circuits are used for processing and interfacing analog signals, the accessibility of these circuits may be drastically low. Built-in self-test is a solution for enhancing testability and fault

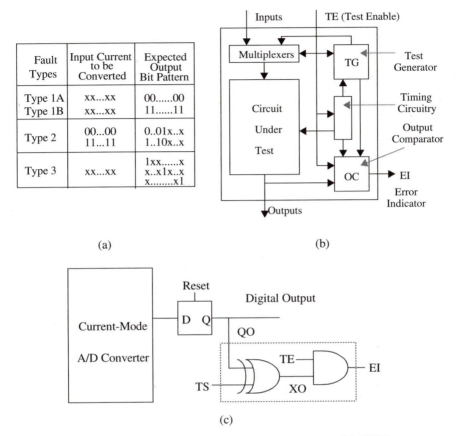

Fault Types	Input Current to be Converted	Expected Output Bit Pattern
Type 1A	xx...xx	00......00
Type 1B	xx...xx	11......11
Type 2	00...00 11...11	0..01x..x 1..10x..x
Type 3	xx...xx	1xx......x x..x1x..x x........x1

(a)

(b)

(c)

Figure 8-7 *Self-test scheme: (a) Test patterns; (b) BIST structure; (c) BIST A/D converter. (Figure courtesy of the IEEE, © 1995 [30], 1996 [44], 1997 [45].)*

diagnosability. The major issues in BIST design include the hardware overhead incurred, the self-test capability of the hardware added, and the performance degradation caused by the additional hardware. For example, in a wide range of schemes, extra test switches have been added to digital circuits to enhance their testability. However, the addition of switches to an analog circuit may affect the performance of the circuit. Therefore, reducing the performance impact of additional switches in a BIST scheme is an important design issue.

A typical BIST structure for a digital circuit, as shown in Figure 8-7(b) [45], is comprised of five major parts: the *Test Generator* (TG), *Input multiplexers* (INMUX), the *Unit Under Test (UUT)*, the *Output Comparator*, and *Timing Circuitry*. The INMUX selects the input signals either from the normal input signals during the operation mode, or from the test signal generated from TG during the test mode. The timing circuitry is used to synchronize the entire operation. Two extra pins, *test enable* (TE) and *error indicator* (EI), are needed. Figure 8-7(c) illustrates a BIST design [31] for the current-mode A/D converter circuit in Figure 8-5(b). The test generator produces the test currents and the expected output bit strings; the converter circuit converts the applied test currents and generates a digital output bit string which is stored in a register; the output comparator compares the data stored in the register with the expected output; and, finally, the timing circuitry is used to synchronize the entire operation and the clock pulses required for controlling the switching elements.

8.3.4.1   BIST Structure

The BIST scheme is based on the test generation techniques discussed in Section 8.3.3.2.

The ***test generator*** generates the test currents: $I_{T1} = 0$ and $I_{T2} = I_{ref}$. Since the reference current I_{ref} already exists in the converter circuit, the reference current and the corresponding switch are used as the test current. In other words, the test current $I_{T2} = I_{ref}$ is implemented by copying I_{ref} to the copier P_1 as the current to be converted during the current S/H cycle. On the other hand, the test current $I_{T1} = 0$ is realized by setting the input current I_{IN} to a zero current and copying it to the copier P_1 as the current to be converted during the S/H cycle. Therefore, no extra hardware is required for generating the test currents. When the test current $I_{T1} = 0$ is applied to the fault-free converter, a bit string of zeros is expected. Thus, a signal TS=0 is generated to represent the expected output bit. Similarly, for $I_{T2} = I_{ref}$, a bit string of ones is expected and the signal TS is set to 1. Therefore, the test generator needs a simple circuit to generate the signal TS.

The ***converter circuit*** converts the test current and produces one bit at a time in each conversion cycle. The converted bit is stored in a *D-type flip-flop*. We assume that the flip-flop has a reset function which allows its content to be reset to 0. The converter circuit samples the input current during the normal operation mode and sets its input current to a zero during the test mode.

A ***1-bit comparator*** is used to compare the data held in the flip-flop with the expected output signal TS, where TS = 1 when the test current $I_{T2} = I_{ref}$ is applied and TS = 0 when $I_{T1} = 0$ is applied. The 1-bit comparator can be realized by a simple *XOR* gate. In order to make the error indication (EI) signal testable, an *AND* gate which takes two inputs, XO and TE, is used. During the normal operation mode, TE is set to 0 and thus EI is also 0. On the other hand, during the test mode, TE = 1 and any mismatches in the comparison results in XO = 1 causing EI to be equal to 1 indicating the presence of an error.

8.3.4.2   Self-Testability

A fault that occurs in any component in the BIST structure causes an error and its presence is indicated by the error signal EI. By faults, we mean the stuck-at ON/ OFF faults at all switching elements in the converter circuit and the stuck-at 0/1 faults at the D-input and Q-output (denoted by QO) of the flip-flop, TS, XO, TE, and EI. The self-testing process starts with a check of the circuitry added for BIST, and then checks the converter circuit for faults.

The circuitry external to the converter is tested by turning on the power and resetting the flip-flop to 0. The test-enable signal TE is changed from 0 to 1, and it is expected that, for a fault-free *AND* gate, the signal EI will also change from 0 to 1, while the signal TS is set to 1. Consequently, an unchanged signal EI = 1 implies that a stuck-at-1 (s-a-1) fault occurs at EI or TE. On the other hand, an unchanged signal EI = 0 indicates the occurrence of either a s-a-0 fault at TE, EI, TS, or XO, or a s-a-1 at QO. Note that the s-a-1 fault at QO also implies that the flip-flop fails to perform the reset function. After passing the above test, the signal TS is set to a 0 and the signal TE remains at 1, hence, the signal EI is expected to be at 0. Therefore, an unexpected EI=1 implies that a s-a-1 fault occurs at TS, QO, EI, or XO. Therefore, with these two tests, the only undetected stuck-at 0/1 faults are the s-a-0 fault at QO and the s-a-0/1 fault at the D-input of the flip-flop which will be tested for later.

From Figure 8-7(a), the converter circuit is tested by first applying a zero test current, where TS is set to 0, the flip-flop is reset, and TE remains at 1 for the test mode. In this test, the signals EI = 0 and XO = 1 are expected for a fault-free circuit. An unexpected EI = 1 implies the occurrence of a *Type 1B, Type 2,* or *Type 3 fault,* or a s-a-1 fault at the D-input of the flip-flop. After passing the test, the test current I_{ref} is applied, while TS is set to 1, the flip-flop is reset, and TE = 1. This test is also expected to produce EI = 0. An unexpected EI = 1 implies the occurrence of either a *Type 1A fault,* or a s-a-0 fault at QO or D-input of the flip-flop.

The BIST structure has been shown to be fully testable for all irredundant single stuck-at faults.

8.4 Scan Structures: Design for Testability

Conceptually, testability can be considered as the ability to control and observe signals at the circuit (internal) nodes. Given the complex chip in Figure 8-8(a), analog and digital blocks are generally tested separately. Each block is isolated with a scan path to increase both controllability and observability [46,47]. A scan chain using digital shift registers is added to the interface between the ADC/DAC and DSP core, as shown in Figure 8-8(b), where each register is connected to a node to be accessed. The scan chain allows the test data to be simultaneously loaded to the register and sequentially shifted out. Only two additional pins are required for scan-in and scan-out and to ensure that the scan chain is fault-free. The same scan design concept can be extended to analog blocks. Analog shift registers (ASRs),

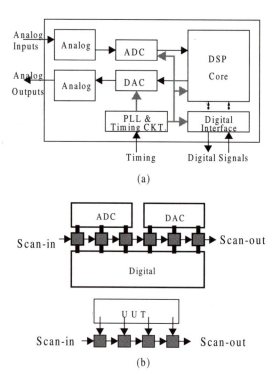

Figure 8-8 *DSP-based mixed-signal IC: (a) Block diagram; (b) Scan structure.*

realized by sample/hold (S/H) circuits, can be used to load and shift out the test data[41-44]. A switched-capacitor (SC) S/H circuit was used to realize the shift register for voltage test data measurement [41]. Current copier-based S/H circuits for current test data measurement in switched-current (SI) circuits are shown in Figure 8-9. The structure in Figure 8-9(a) is a shift register-based structure while the structure in Figure 8-9(b) uses analog multiplexing and demultiplexing on the inputs to and the outputs from the scan path.

8.5 Conclusion

High-performance analog interface and data acquisition/conversion circuits are commonly designed using *switched-capacitor* (SC) or *switched-current* (SI) techniques. Self-calibration, self-correction, and/or self-compensation mechanism(s) are usually employed to eliminate offset and nonlinearities, and to cancel the error effects. Though these mechanism(s) may work properly when some com-

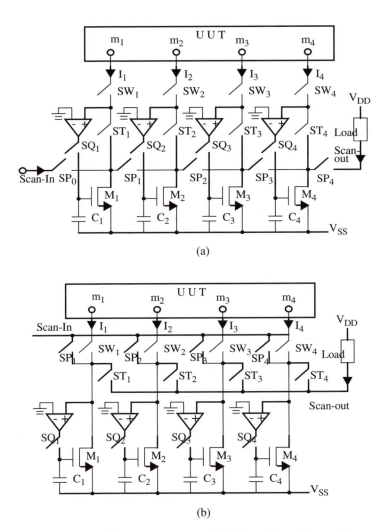

Figure 8-9 *SI scan structures: (a) shift register; (b) mux/de-mux. (Figure courtesy of the IEEE, © 1992 [42].)*

ponent values deviate from their nominal values, the mechanism will no longer function properly in the presence of faulty switching elements. Current is the primary information-carrying medium in the switched-current circuits. Relative to switched capacitor circuits, they offer several advantages. High-performance SI circuits are expected to be intensively used in low-power/low-voltage environment. Testing SI circuits in such an environment is a challenging task.

This chapter reviewed current test methods for switched-current circuits. The review included testable designs for current copiers and filters [37,38], self-testing designs for current copiers and SI building blocks [36], and a built-in current tester design [33,39]. In addition, for the CMOS current-mode A/D converter in [7], its concurrent error capability, test generation [25-30], and built-in self-test (BIST) design [31] are also discussed. Current copiers can also be used as scan registers [41-43] for testability enhancement.

In the future, as integration increases, access to switched current components can be expected to decrease. Thus, design-for-testability and BIST (Built-in Self-test) technologies have been recognized as practical solutions to the test problem. Similarly, accurately measuring in-circuit currents externally may not be an easy task. Therefore, the development of an efficient and effective built-in current tester to create easily testable designs is an important task for future study. A related problem is the development of an effective fault model to analyze test schemes. According to recently developed testability analysis methodologies [48,49] using inductive fault analysis (IFA)[50], the percentage of stuck-at-ON and stuck-at-OFF faults in a CMOS transistor as a switch is approximately 70%. Thus, the defect coverage for fault models which only consider catastrophic faults is insufficient. An efficient and effective fault model for switches is essential for switched-current circuits.

References

1. A. Mastsuzawa, "Low-voltage and low-power circuit design for mixed analog/digital systems in portable equipment," *IEEE Journal of Solid-State Circuits*, vol. 29, no. 4, pp. 470-480, April 1994.
2. C.L. Wey, "Mixed-signal testing: a review," *IEEE International Conference on Electronics, Circuits, and Systems*, Rodos, Greece, pp. 1064-1067, October 1996.
3. C.L. Wey, "Mixed-signal fault models and design for test," *IEEE Asian Test Symposium*, Hsinchu, Taiwan, November 1996.
4. R. Gregorian and G.C. Temes, *Analog MOS Integrated Circuits for Signal Processing*, New York: John Wiley & Sons, 1986.
5. C. Toumazou, J.B. Hughes, and N.C. Battersby, *Switched-Currents: An Analogue Technique for Digital Technology*, London: Peter Peregrinus Ltd, 1993.
6. S.J. Daubert, D. Vallancourt, and Y.P. Tsivids, "Current copier cells," *Electronics Letters*, vol. 24, no. 25, pp. 1560-1562, December 1988.
7. D.G. Nairn and C.S. Salama, "A ratio-independent algorithmic analog-to-digital converter combining current mode and dynamic techniques," *IEEE Trans. on Circuits and Systems*, vol. 37, no.3, pp. 319-325, March 1990.
8. C.L. Wey and S. Krishnan, "An accurate current-mode divide-by-two circuit," *Electronics Letters*, vol. 28, no. 9, pp. 820-822, April 1992.
9. J.B. Hughs and K.W. Moulding, "Switched-current signal processing for video frequencies and beyond," *IEEE Journal of Solid-State Circuits*, vol. 28, no. 3, pp. 314-318, March 1993.
10. R.H. Zele and D.J. Allstot, "Low-voltage fully-differential CMOS switched-current filters," *Proc. of IEEE Custom Integrated Circuits Conference*, pp. 6.2.1-6.2.4, 1993.
11. J.B. Hughs and K.W. Moulding, "S^2I: A switched-current technique for high performance," *Elec-*

tronics Letters, vol. 29, no. 16, pp. 1400-1401, August 1993.

12.   S. Krishnan and C.L. Wey, "An accurate reference-generating circuit for successive approximation current-mode a/d converters," *International Journal of Circuit Theory and Application*, no. 21, pp. 361-369, August 1993.

13.   I. Mehr and T. Sculley, "A 16-bit current sample/hold circuit using a digital CMOS process," *Proc. International Symp. on Circuits and Sys.*, London, pp. 417-420, May 1994.

14.   S. Krishnan and C.L. Wey, "A parallel current-mode a/d converter array with a common current reference-generating circuit," *Proc. 37th Midwest Symp. on Circuits and Systems*, Lafayette, LA, pp. 1168-1171, August 1994.

15.   D. Macq and P.G.A. Jespers, "A 10-Bit pipelined switched current A/D converter," *IEEE Journal of Solid-State Circuits*, vol. 29. no. 8, pp. 967-971, August 1994.

16.   C.-Y. Wu, C.-C. Chen, and J.-J. Cho, "Precise CMOS current sample/hold circuits using differential clock feedthrough attenuation techniques," *IEEE Journal of Solid-State Circuits*, vol. 30, no.1, pp. 76-81. January 1995.

17.   R. Huang and C.L. Wey, "Simple yet accurate current copiers for low-voltage current-mode signal processing applications," *International Journal of Circuit Theory and Application*, vol. 23, pp. 137-145, March 1995.

18.   R. Huang and C.L. Wey, "Simple low-voltage, high-speed, high-linearity v-i converter with s/h for analog signal processing applications," *IEEE Trans. on Circuits and Systems Part II*: *Analog and Digital Signal Processing*. vol. 143, no. 1, pp. 52-55, January 1996.

19.   R. Huang and C.L. Wey, "Design of high-speed, high-accuracy current copiers for low-voltage analog signal processing applications," *IEEE Trans. on Circuits and Systems Part II*: *Analog and Digital Signal Processing*, vol.43, no.12, pp. 836-839, December 1996

20.   R. Huang and C.L. Wey, "A 5mW, 12-b, 50ns/b switched-current cyclic A/D converter," *Proc. IEEE International Symposium on Circuits and Systems*, Atlanta, GA, vol. I, pp. 207-210, May 1996.

21.   R. Huang and C.L. Wey, "A high-accuracy CMOS oversampling current sample/hold (S/H) circuit using feedforward approach," *Proc. IEEE International Symp. on Circuits and Systems*, Atlanta, GA, vol. I, pp. 65-68, May 1996.

22.   R. Huang, High-performance CMOS Switched-Current Circuits for Low-Voltage Analog Signal Processing Applications, *Ph.D. Dissertation*, Department of Electrical Engineering, Michigan State University, August 1997.

23.   Y.P. Tisividis, "Analog MOS integrated circuits - certain new ideas, trends, and obstacles," *IEEE Journal of Solid-State Circuits*, vol. SC-22, pp. 317-321, June 1987.

24.   H.T. Yung and K.S. Chao, "An error-compensation A/D conversion technique," *IEEE Trans. on Circuits and Systems*, vol. 38, pp. 187-195, February 1991.

25.   C.L. Wey, "Concurrent error detection in current-mode A/D converter," *Electronics Letters*, vol. 27, no. 25, pp. 2370-2372, December 1991.

26.   S. Sahli, Test Generation and Concurrent Error Detection in Current-Mode A/D Converters, *M.S. Thesis*, Department of Electrical Engineering, Michigan State University, May 1992.

27.   S. Krishnan, S., Sahli, and C.L. Wey, "Test generation and concurrent error detection in current-mode A/D converters," *Proc. IEEE International Test Conference*, Baltimore, MD., pp. 312-320, September 1992.

28.   C.L. Wey, S. Krishnan, and S. Sahli, "Design of concurrent error detectable current-mode A/D converters for real-time applications," *Analog Integrated Circuits and Signal Processing*, no. 4, pp. 65-74, July 1993.

29.   S. Krishnan, Design and Test of Current-Mode Signal Processing Circuits, *Ph.D. Dissertation*, Department of Electrical Engineering, Michigan State University, August 1993.

30. C.L. Wey, S. Krishnan, and S. Sahli, "Test generation and concurrent error detection in current-mode A/D converters," *IEEE Trans. on CAD*, vol. 14, no.10, pp. 1191-1198, October 1995.

31. C.L. Wey, "Built-in self-test (BIST) design of current-mode algorithmic A/D converter" *IEEE Trans. on Instrumentation and Measurement*, vol. IM-46, no. 3, pp. 667-671, June 1997.

32. C.-P. Wang and C.L. Wey, "Test generation of switched-current A/D converter," *Proc. 2nd IEEE International Mixed signal Testing Workshop*, Quebec City, Canada, pp. 98-103, May 1996.

33. C.-P. Wang and C.L. Wey, "Test generation of analog switched-current circuits," *Proc. Asian Test Symposiums*, Taiwan, pp. 376-381, November 1996.

34. C.-P. Wang and C.L. Wey, "Fault macromodel for switches in switched-current circuits," to appear in *International Journal of Circuit Theory and Application*.

35. C.-P. Wang, Efficient Testability Design Methodologies for Mixed-Signal/Analog Integrated Circuits, *Ph.D. Dissertation*, Department of Electrical Engineering, Michigan State University, August 1997.

36. T. Olbrich and A. Richardson, "Design and self-test for switched-current building blocks," *IEEE Design & Test of Computers*, pp. 10-17, Summer 1996.

37. G. Taylor, P. Wrighton, C. Toumazou, and N. Battersby, "Mixed-signal testing considerations for switched current signal processing," *Proc. Midwest Symposiums on Circuits and Systems*, Washington D.C., August 1992.

38. P. Wrighton, G. Taylor, I. Bell. and C. Toumazou, "Test for switched-current circuits," in *Switched-Currents: An Analogue Technique for Digital Technology*, edited by C. Toumazou, J.B. Hughes, and N.C. Battersby, London: Peter Peregrinus Ltd, pp.487-507, 1993.

39. C.-P. Wang and C.L. Wey, "High-accurate cmos current comparator," *Proc. Midwest Symposiums on Circuits and Systems*, Davis, CA, August 1997.

40. P.M. Dias, J.E. Franca, and N. Paulino, "Oscillation test methodology for digitally-programmable switched-current BIQUAD," *IEEE International Mixed signal Testing Workshop*, Quebec City, Canada, pp. 221-226, May 1996.

41. C.L. Wey, "Alternative built-in self-test structure (BIST) for analog circuit fault diagnosis," *Electronics Letters*, vol. 27, no. 18, pp. 1627-1628, August 1991.

42. C.L. Wey and S. Krishnan, "Built-in self-test (BIST) structures for analog circuit fault diagnosis with current test data," *IEEE Trans. on Instrumentation and Measurement*, vol. IM-41, no.4, pp. 535-539, August 1992.

43. M. Soma, "Structure and concepts for current-based analog scan," *Proc. Custom Integrated-Circuits Conference*, pp. 517-520, 1995.

44. C.L. Wey, "Built-in self-test (BIST) structure for analog circuits fault diagnosis," *IEEE Trans. on Instrumentation and Measurement*, vol. IM-39, no.2, pp. 517-521, June 1990.

45. C.L. Wey, "Built-in self-test (BIST) design of high-speed carry-free dividers," *IEEE Trans. on VLSI Systems*, vol. 4, no. 1, pp.141-145, March 1996.

46. P.P. Fasang, "Analog/digital ASIC design for testability", *IEEE Trans. on Industrial Electronics*, vol. 36, no. 2, pp. 219-226, May 1989.

47. K.D. Wagner and T.W. Williams, "Design for testability of analog/digital networks," *IEEE Trans. on Industrial Electronics*, vol. 36, no. 2, pp. 227-230, May 1989.

48. C.-P. Wang and C.L. Wey, "Efficient testability design methodologies for mixed-signal/analog integrated circuits," *Proc. 3rd IEEE International Mixed Signal Testing Workshop*, Seattle, WA, pp. 68-74, June 1997.

49. C.-P. Wang and C.L. Wey, "Development of hierarchical testability design methodologies for mixed-signal/analog integrated circuits," *Proc. International Conference on Computer Design*, pp. 468-474, October 1997.

50. J.P. Shen, W. Maly, and F.J. Ferguson, "Inductive fault analysis of MOS integrated circuits," *IEEE*

Design and Test of Computers, pp. 13-26, December 1985.

Index

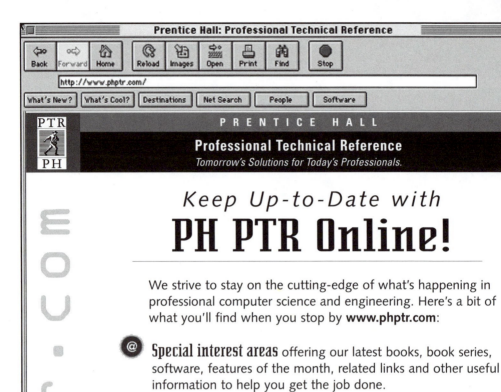